Aviation is not inherently unsafe,
but like the sea, it is terribly
unforgiving of any carelessness or neglect.

SUNY Series, Case Studies in Applied Ethics, Technology, and Society
John H. Fielder, editor

The DC-10 Case

A Study in Applied Ethics, Technology, and Society

John H. Fielder
and
Douglas Birsch,
Editors

STATE UNIVERSITY OF NEW YORK PRESS

Published by
State University of New York Press, Albany

For information, address State University of New York Press,
State University Plaza, Albany, N.Y., 12246

Production by Marilyn P. Semerad
Marketing by Fran Keneston

Library of Congress Cataloging-in-Publication Data

The DC-10 case : a study in applied ethics, technology, and society /
    edited by John H. Fielder and Douglas Birsch.
        p.   cm. — (SUNY series, case studies in applied ethics,
    technology, and society)
    Includes bibliographical references.
    ISBN 0-7914-1087-0 (hardcover). — ISBN 0-7914-1088-9 (paper)
    1. McDonnell Douglas DC-10 (Jet transport) 2. Aeronautics-
-Accidents. 3. Airplanes—Crashworthiness. 4. Jet transports-
-Design and construction. 5. Aircraft industry—Corrupt practices.
I. Fielder, John H., 1939-   . II. Birsch, Douglas, 1951-   .
III. Series.
TL686.M25D3   1992
363.12′416—dc20   91-28585
                                                            CIP

10 9 8 7 6 5 4 3 2 1

For Pat Lawler
and
Ellen, Kathryn, and Jocelyn Birsch

Contents

THE AVIATION SAFETY SYSTEM

Preface

The DC-10 is a wide-body jumbo jet manufactured by McDonnell Douglas in Long Beach, California, a suburb of Los Angeles. It is powered by three high-bypass fan jet engines, one under each wing and one in the tail fin just above the fuselage. It is the engine mounted high on the tail that provides the most characteristic visual feature of the DC-10. Like its U.S. competitors, the Boeing 747 and the Lockheed L-1011 TriStar, it is capable of carrying several hundred passengers or very large cargo loads for thousands of miles. It was introduced into commercial service in 1972 with the usual combination of pride and promotion that accompany such events.

Unfortunately the DC-10's history has been marked by a series of crashes which have convinced many observers that the airplane contained serious design weaknesses. Concern about the DC-10's safety centers on three accidents, Paris (1974), Chicago (1979), and Sioux City (1989), which took 730 lives. In each case failures that should not have led to a crash—the opening of a cargo door, a wing engine falling off, and explosive disintegration of the tail engine—severely damaged the control system of the DC-10. It was this secondary damage that caused the crashes. Critics regard the vulnerability of the DC-10's control system as its deadly design flaw.

These events raised the central question of the DC-10 case: was the design of the aircraft adequately safe? More specifically, was its control system adequately protected from secondary damage? From an ethical standpoint this question may be stated as: did McDonnell Douglas meet its ethical obligations to design and build an aircraft that met appropriate safety standards? This book exists because a substantial number of people believe that the answer to that question is no.

If the DC-10 was allowed to enter commercial service with serious safety problems, a series of ethical questions arise concerning the actions of the organizations that built it (McDonnell Douglas and

Convair, a subcontractor), individuals in those organizations who knew about the safety problems of the DC-10, and the policies, performance, and institutional arrangements of the aviation safety system and specifically of the Federal Aviation Administration. Were their responses to problems with the safety of the DC-10 ethically acceptable? Did they meet their obligations to the flying public? Are the policies and institutions of the aviation safety system ethically acceptable? Do they meet the ethical standards we require for the protection of airplane passengers?

An examination of these questions leads us into a complex and fascinating story of the way airplanes are designed, financed, produced, sold, regulated, investigated, and litigated. At every step of this process there are ethical choices to be made. It is tempting to condemn those whose actions (and inactions) failed to prevent hundreds of deaths, but any judgment about ethical failure must take into consideration the actual circumstances in which choices had to be made. A fair ethical assessment must take into account the options available, the risks to individuals and organizations associated with particular responses, decisions made under uncertainty, and the web of organizational and institutional constraints that surrounded every choice.

USING THIS BOOK

Making ethical judgments about the DC-10 case requires a reflective awareness of the basic ethical concepts we use to analyze ethical questions, some knowledge of the complex web of circumstances that shaped decisions and policies, and an acquaintance with what has been written about ethical issues in the DC-10 case. Each of these topics is addressed in this book.

The Introduction summarizes the basic facts and ethical issues of the case and gives a brief description of the chapters, which provide additional factual detail or explore the ethical issues in greater depth. The former include excerpts from official accident reports, discussions of the aircraft manufacturing industry, explanations of technical features of the DC-10, and additional information about the persons, organizations, and events that contributed to the story. The latter consist of articles and essays on ethical issues in the DC-10 case, such as the professional and moral responsibilities of engineers, whistleblowing, the ethical obligations of organizations, and public policy questions concerning aviation safety. These contributions invite the reader to agree or disagree with their ethical assessments of persons, policies, organizations, and institutional arrangements.

"Ethical Analysis of Case Studies" is an introduction to the basic concepts of ethical analysis. Confusion is as great an enemy of thought as ignorance, and this essay is included to clarify and make explicit the ideas we use in ethical assessment. It provides a vocabulary with which the ethical issues in the DC-10 case can be more effectively articulated and discussed.

The DC-10 case presents us with a smorgasbord of ethical issues, policy questions, and institutional problems. It can be sampled in many different ways, tracing different themes through the sequence of events. We have focused on ethical responsibility, but the case is also an excellent opportunity to raise questions about the adequacy of government regulation in a capitalist society, managerial decision making, and the legal and financial relationships among large corporations.

Finally, the DC-10 case is simply a fascinating story in its own right. Decisions were made against a background of financial crisis, conflicting bureaucratic objectives, and the technical problems of an enormous, complex machine. And always, in the background, the knowledge that hundreds of innocent persons would die if those decisions were wrong.

Many people contributed to this book in various ways. We wish to give special thanks to the Aviation Safety Advisory Panel of the Airline Passengers Association of North America, particularly to Dick Livingston, Bill Jennings, Tom Lombardo, and Norman Birch. At Villanova University, Sandy Shupard, Mary Quilter, Jack Doody, Charles Marston, Don Joye, Lee Christensen, Bob Lynch, and Fr. Lawrence Gallen have all made special contributions to our efforts. Michael Gathje kindly performed the onerous task of preparing the index. Richard Lipow provided advice and encouragement, and Carola Sautter and Marilyn Semerad at SUNY guided us through the publishing process and were always available to give advice and suggestions. Most of all, our wives, Pat Lawler and Ellen Birsch, have sustained us through many difficulties with their ideas, enthusiasm and emotional support.

Introduction

The DC-10 case is a complex story of persons, machines, accidents, laws, policies, organizations, and institutions. An acquaintance with the major features of that story is essential for an ethical evaluation of key policies and actions. This introduction provides an outline of the events and issues in the case and brief descriptions of the articles.

The selections that follow develop specific parts of the story in more detail or examine the ethical responsibilities of individuals, corporations, and the Federal Aviation Administration (FAA). The former selections include excerpts from government investigations of DC10 crashes, and studies of the DC-10 and of the aviation industry. Each develops an important part of the story, describing accidents, DC-10 design features, policies, decisions, and events. The latter include articles on the professional and moral responsibilities of engineers, whistleblowing and ethical obligation, and critiques of the FAA and the aviation industry. Together they provide the material needed for a detailed ethical analysis of the DC-10 case.

HISTORY AND EARLY WARNINGS

The history of the aviation industry really begins with the establishment of the U.S. Air Mail Service in the early 1920s. Companies were established to provide mail service and many of them grew into today's airlines. Federal regulation of safety also began in that period, and its structure and orientation reflect the historical events that shaped the industry and its regulation. "Regulatory and Institutional Framework" traces the history of federal regulation of air safety and outlines some of the ongoing concerns about its ability to carry out its responsibilities.

The Douglas Aircraft Company was the leading U.S. aircraft

manufacturer for many years, but it was late in making the transition from propeller to jet aircraft, allowing Boeing to take the lead first with its 707 and later with the popular 727. In an attempt to attract buyers, Douglas offered many customized versions of its DC-8 and DC-9 jet aircraft. This customizing of the production process caused problems which were compounded by shortages of technically proficient employees and critical materials during the Vietnam war. In 1967 Douglas was experiencing severe financial difficulties and was purchased by the McDonnell Corporation, a manufacturer of military aircraft based in St. Louis. Douglas became a division of McDonnell Douglas.

Still trying to catch up with Boeing, the new company belatedly entered the jumbo jet market, where Boeing again held the lead with its highly successful 747. McDonnell Douglas found itself in a cruel race for second place with Lockheed, which was re-entering the commercial market with a similar wide-body jet with three engines. Because the market at that time would not support three manufacturers of jumbo jets, there would be no third place. Thus the DC-10 began its existence in an atmosphere of change and urgency.

In the aircraft business the financial risks and corresponding stresses are very high. The commitment to build a plane like the DC-10 requires virtually all of the company's net worth: they call it "betting the company." Lockheed lost $2.5 billion before it stopped making its jumbo jet, the L-101 1, and withdrew entirely from the commercial aircraft business. Boeing, in the words of one of its managers, was a "gnat's whisker" away from bankruptcy at one point in the development of its famous 747. Douglas has not yet reached the break-even point for its DC-10, and may never do so. When survival cannot be taken for granted, the pressures on individuals, corporations, and oversight agencies are immense and pervasive. All are aware of playing in a high-stakes game and respond accordingly. John Newhouse describes this dimension of the aircraft business in "High Risks, Sinking Fortunes."

Even before the DC-10 had its first test flight, there were warnings of design problems that would later claim many lives. The first DC-10 produced (Ship l) was being tested on the ground in 1970 when the forward cargo door blew open, causing part of the floor in the passenger cabin to collapse. Because the floors of jumbo jets are designed to function with roughly equal air pressure from the passenger cabin above and the cargo hold below, sudden depressurization of either (which would result from a door opening) will cause the floor to buckle. Control systems that actuate movements of flying surfaces run through the floor, and any damage to it threatens the pilot's ability to fly the aircraft. John Fielder's chapter, "Floors, Doors, Latches, and Locks," explains the design of the floor and the cargo doors of the DC-10 and shows why the airplane was especially vulnerable to loss of control resulting from inadvertent opening of the cargo door in flight.

The discovery of this weakness in 1970 led to revisions of the cargo doors and created a financial and design dispute between Convair, a subcontractor, and McDonnell Douglas over what changes should be made and who should pay for them. "The 1970 Ground Testing Incident" details the concerns of individuals in the Convair organization about the door and how the legal relationships among McDonnell Douglas, Convair, and the FAA complicated their response to this difficulty. Although the problem was being addressed by the manufacturers, a complex network of financial, technical, and legal considerations made effective action more difficult.

A number of ethical issues emerge here concerning the policies of the FAA. Before an aircraft can receive a certificate of airworthiness, its legal right to be flown, it must undergo many inspections, tests, and analyses. For financial and proprietary reasons, most of those inspections are done by the manufacturer's employees called "designated engineering representatives" (DERs). They make inspections for the FAA on their own aircraft, a clear conflict of interest. Many persons have claimed that this arrangement and other FAA policies are ethically doubtful if not unacceptable.

In June 1972, less than a month after the DC-10 had been put into service, American Airlines Flight 96 out of Detroit suffered loss of a rear cargo door over Windsor, Ontario. Because the plane was only lightly loaded (only 67 passengers and crew in an airplane that can hold 350), there was only a partial collapse of the floor. Some control lines running through it were rendered inoperable, but owing to the skill of the pilot, the damaged aircraft was brought down in a harrowing but safe landing. "National Transportation Safety Board Report on the Windsor Incident" analyzes the control system damage and reveals how the design changes initiated after the 1970 ground test accident were inadequate to protect the aircraft. It also suggests that the DC-10 should not have been certificated with a control system that could be disabled by the loss of a door. Critics believe that the FAA did not meet its ethical obligation to make sure that the DC-10 was adequately safe.

Shortly after the Windsor incident, Dan Applegate, director of product engineering for Convair's part of the DC-10 project, wrote a memo which clearly set out the design faults of the cargo door latch and lock system and urged his superior to take action. It is one of the most powerful documents in this case study, clearly describing his grave doubts about the safety of the aircraft. Applegate's immediate superiors, J. B. Hurt and M. C. Curtis, decided not to act on Applegate's recommendation about a solution to the problem with the cargo door. When Hurt received Applegate's memo, McDonnell Douglas and its subcontractor Convair were already disputing over who should pay for earlier fixes to the plane. By contract, Convair was forbidden from contacting the FAA directly about problems with the aircraft. The only legal option open to them was to recommend a better fix of the problem to McDonnell Douglas. Hurt and Curtis may have felt that Convair's recommendation of a radical fix to Douglas might lead to Convair being held liable for paying for that alteration. For whatever reason, they failed to act on Applegate's memo and he apparently accepted their decision and did not pursue the matter. Applegate's memorandum and Convair's response to it are described in "The Applegate Memorandum," which further explains why neither Dan Applegate nor Convair took action in response to this remarkable document. Whether they were ethically obligated to take this matter further is one of the central questions in the DC-10 case.

When there is a safety problem with an airplane, the FAA typically issues an airworthiness directive (AD), which requires corrective action to be taken on the aircraft within a specific time. Airworthiness directives have the force of law; they must be carried out in order to continue legal operation of the plane. They are also public documents, and issuance of a major AD after Windsor would have been an embarrassment to McDonnell Douglas, who was actively seeking buyers for its new DC-10. John Shaffer, head of the FAA, and Jackson McGowen, President of the Douglas Division of McDonnell Douglas, reached a "gentleman's agreement" that the plane would be fixed by service bulletins to the airlines issued by McDonnell Douglas. "Fat, Dumb, and Happy: The Failure of the FAA" traces the events that led to this solution and its consequences. Shaffer has been criticized for failing to issue an AD at this time. But it can be argued that Shaffer was conscientiously carrying out the objectives of the FAA charter, which requires that the FAA both promote the aviation industry and provide for its safety. This is an obvious source of conflict, and many believe that it is ethically unacceptable to have an institution with these two roles.

The use of service bulletins to fix the cargo door did not generate rapid compliance throughout the airline industry. "Compliance
with service Bulletin SB 52-37" shows that many aircraft were not
modified until the following year, and one had still not received any
changes when the 1974 Paris crash occurred. Congressional investigators were also critical of the gentleman's agreement, as "Conclusions of the U.S. Senate Oversight Hearings and Investigation of
the DC-10 Aircraft" reveals.

THE 1974 PARIS CRASH

The modifications required in the service bulletins issued after
the Windsor incident were not all carried out on Ship 29, which was
still in McDonnell Douglas's possession. Forged or mistaken records
exist to show that the required modifications were made, but in fact
only one was completed, the addition a small viewing port to allow
visual inspection of a lock pin. Ship 29 was later sold to Turkish
Airlines, Turk Hava Yollari (THY), with only a partially modified
rear cargo door. In March, 1974, it crashed outside of Paris, killing
all 346 people on board. Investigators found six bodies and the rear
cargo door several miles from the crash site. It was a virtual repeat
of the Windsor incident, only this time with a fully loaded aircraft.
"The French Report on the 1974 Paris Crash" provides chilling
details of Ship 29's last flight.

Following the crash, the FAA issued an airworthiness directive to inspect the DC-10 fleet and make sure that modifications to
the cargo door set out in earlier service bulletins had been carried
out. A year later another AD required that the floors of all jumbo jets
be able to withstand a sudden opening in the hull of 20 square feet
(the rear cargo door is 14.5 square feet). Manufacturers met this
requirement through a combination of stronger floors and the addition of pressure relief vents in the floors, which would allow excess
air pressure to flow through the floor without causing it to collapse.

The Paris crash raises a host of ethical and professional issues.
Applegate, Hurt, and Curtis all knew of the door problem and could
have contacted McDonnell Douglas about it. Should these three individuals bear any of the ethical responsibility for the subsequent
crash near Paris and the loss of 346 lives? Does Applegate have a
special responsibility for the crash because, as an engineer, he was in
the best position to understand the technical problems involved?

Since Hurt and Curtis were acting as agents of Convair, does the company bear any responsibility for the Paris crash?

Did Dan Applegate live up to his professional and ethical responsibilities in connection with the DC-10? In his chapter, "Engineers Who Kill: Professional Ethics and the Paramountcy of Public Safety," Kenneth Kipnis argues that professionalism involves more than simply expertise; it involves a public commitment to some significant set of social values. He believes that engineers should not participate in projects that degrade existing ambient levels of public safety unless information concerning those degradations is made generally available. Kipnis discusses the DC-10 case and Applegate's role in it and concludes that Applegate should have been aware that the use of the DC-10 aircraft represented a substantial degradation of the existing ambient level of safety for airline passengers. Applegate should also have realized that adequate information about the problem with the DC-10 was not being provided to the FAA. According to Kipnis, Dan Applegate knew that he was involved in a project that degraded ambient levels of public safety without adequate information about that degradation being generally available. Thus, he must share the blame for the crash of the Turkish Airlines DC-10.

The chapter by Douglas Birsch, "Whistleblowing, Ethical Obligation, and the DC-10 Case," examines the question of whether Dan Applegate should have blown the whistle on Convair and McDonnell Douglas. Birsch presents two approaches to whistleblowing and two conclusions about Applegate. The first approach, developed by Richard DeGeorge, presents criteria which can be used to determine when whistleblowing is morally permissible and morally obligatory. Based on DeGeorge's conditions, Birsch concludes that it would have been permissible but not obligatory for Applegate to blow the whistle. Birsch criticizes DeGeorge's conditions by arguing that they have two main faults: they are too vague to make them practical and they allow people to escape from being involuntarily obligated to blow the whistle. This lack of involuntary obligation distorts the usual notion of ethical obligation. Birsch contrasts DeGeorge's position on whistleblowing with his own view which suggests that Applegate was ethically obligated to blow the whistle on Convair. He argues that his position is superior to DeGeorge's because it is more consistent with our usual view of ethical obligation and because it would be more likely to prevent tragedies like the Paris crash of 1974.

An interesting feature of the literature on the DC-10 case is that it tends to focus on Dan Applegate, as in the chapter by Kipnis. It should be remembered that there are other engineers at Convair and at McDonnell Douglas who are just as accountable as Dan Applegate, but there is no public knowledge of their actions.

McDonnell Douglas's response to the door problem raises as many questions as Convair's involvement. Douglas engineers were aware of the door problem at least as early as the 1970 test failure on Ship 1. While Convair was forbidden from contacting the FAA about the problem, there were no such legal limitations on Douglas. McDonnell Douglas continually rejected the suggestions of the Convair engineers, such as latches driven by a hydraulic system and blow-out panels in the cabin floor, that would have resulted in a safer plane. In addition, Douglas handled the door repair, required by the gentleman's agreement after the Windsor incident, by issuing maintenance bulletins. This allowed dangerous planes to keep flying since they were not legally required to be fixed, nor were carriers adequately notified of the seriousness of the problem. Should the engineers and executives at Douglas bear any responsibility for the Paris crash?

Perhaps the most tragic of McDonnell Douglas's shortcomings was that their maintenance procedure allowed Ship 29, the aircraft that crashed near Paris, to enter into service with certification stamps stating that it had the required changes when those repairs had never been done. Should McDonnell Douglas bear the primary responsibility for the crash of Ship 29 since they controlled the DC-10 project and were responsible for failing to upgrade the door on Ship 29? Does the fact that Convair was Douglas's subcontractor and worked under its direction shift the burden of blame from Convair and Applegate to Douglas?

Peter French's chapter "What is Hamlet to McDonnell Douglas or McDonnell Douglas to Hamlet?: DC-10" moves the discussion from Convair and Dan Applegate to McDonnell Douglas by examining the issue of whether McDonnell Douglas is morally responsible for the crash near Paris and the death of 346 people. He argues that engineers and managers in the corporation designed an airplane which they should have known did not meet the engineering standards of the industry with respect to certain vital systems. While they did not intend for the plane to crash near Paris, they knew that they were manufacturing a plane with a higher risk of crashing than their competitors. Not only did they know of the shortcomings of their design, they were willing to manufacture and market this aircraft. French concludes that McDonnell Douglas can be held morally accountable for the Paris crash. In his commentary, Homer Sewell disagrees with some of French's criticisms of McDonnell Douglas and with his call for a drastic overhaul of their quality control system. The "Statement of John C. Brizendine, President, Douglas Aircraft Company, McDonnell Douglas Corporation" presents a brief defense of his company, presented in a congressional hearing following the crash.

THE 1979 CHICAGO CRASH

While there have been no more accidents caused by cargo doors, other accidents have also cast doubt on the safety of the DC-10's control system. In 1979 an American Airlines flight out of Chicago crashed on takeoff when its left engine broke loose and accelerated upward, severing control cables and hydraulic lines in the leading edge of the wing. This caused an uncommanded retraction of the slats, movable extensions of the wing on its leading edge which are deployed to provide additional lift at takeoff and landing. When its slats retracted, the wing lost lift while the other did not. Because of the relatively low speed of the aircraft, the damaged wing stalled, i.e. lost its ability to provide lift. It dipped as the other wing rose, until the wings were perpendicular to the ground and the plane crashed. There were 273 deaths in this accident.

The immediate reason for the loss of the engine was an improper maintenance procedure by the airline, which had caused cracks in the huge pylon holding the engine to the wing. "The National Transportation Safety Board Report on the 1979 Chicago Crash" explains the FAA maintenance policies and raises questions about the vulnerability of the slat control system and the lack of warning to the pilot. (Had the pilot known what was happening, he could have prevented the accident.) Additional questions arose when it was learned that McDonnell Douglas knew about the improper maintenance techniques that led to the pylon cracks. Besides questions about the ethical adequacy of the FAA's maintenance policies, should McDonnell Douglas have informed the FAA that airlines were using a nonstandard method for removing the engine and pylon assembly?

In response to growing public concern about the DC-10, McDonnell Douglas issued "The DC-10: A Special Report" in 1980 to explain its side of the story. It poses a number of questions that have been raised about the pylon that failed, the safety of the DC-10's slat control system, and hydraulic lines. It clarifies many technical issues and ends with comparative data concerning the safety of the DC-10 and other wide-body jets.

The section on the Chicago crash ends with the chapter by Martin Curd and Larry May, "Two Models of Professional Responsibility." Curd and May set out two models of professional responsibility and apply these models to the Paris crash and the Chicago crash. Based on the first model, the malpractice model of responsibility, Curd and May believe that both Convair and McDonnell Douglas management and the engineers who worked on the DC-10, including Applegate, bear responsibility for the Paris crash of 1974. They apply the second model, the reasonable care model of professional responsibility, to the 1979 Chicago crash. They find that design engineers, McDonnell Douglas management, and American Airlines were responsible for that crash. In general, Curd and May believe that professional engineers must do more than write memos when corporations put profits ahead of safety. The engineer is the one most likely to know the dangers of highly technical products, like aircraft, and should accept responsibility for ensuring that the design is adequately safe.

THE 1989 SIOUX CITY CRASH

Debate continues today over the DC-10's safety because of the 1989 crash of United Airlines Flight 232 in Sioux City, Iowa, in which 111 people died. Questions have again been raised about the adequacy of protection for the control systems in the DC-10. The "National Transportation Safety Board Report on the 1989 Sioux City Crash" found that the cause of the accident was a metallurgical flaw in the fan assembly of the rear engine. A crack in the 370-pound fan disk started from the flaw and eventually caused it to disintegrate. Debris from the explosion severed all hydraulic lines, depriving the pilot of control over the flying surfaces on the tail and wings. The National Transportation Safety Board (NTSB) report also describes the weaknesses of the inspection system that is supposed to find flaws before they can cause damage, and it questions the certification of an aircraft with little protection of its hydraulic lines from uncontained engine failure debris.

John Fielder's "The 1989 Sioux City Crash" discusses the ethical issues of a tail engine design and compares the DC-10 with other aircraft. As in the other accidents, loss of the engine should not have resulted in loss of the aircraft but, as in the other accidents, subsequent damage to the DC-10's control system led to the crash. The DC-10 has three separate hydraulic systems, but all three converge close together near the tail. As a result, shrapnel from an exploding engine can disable all three at once, leaving the plane without hydraulic control. An AD has been issued that requires the installation of devices to one hydraulic system which will prevent complete loss of hydraulic fluid in the rest of the system if it is damaged in the tail. An important ethical issue in this accident is the lack of initial protection for the hydraulic lines in the original design. Another is the familiar tradeoff between safety and economy that is played out in the role of testing and inspection in the aviation regulatory system.

THE AVIATION SAFETY SYSTEM

The final part of the book deals with concerns about the institutional system set up to provide for aviation safety. The gentleman's agreement between McDonnell Douglas and the FAA, and the failure of the FAA to provide sufficient oversight of maintenance procedures to prevent the Chicago crash raise questions about the responsibilities of the FAA and whether it is doing an adequate job of fulfilling them. Ralph Nader's testimony before a subcommittee of the House of Representatives explores whether the FAA is adequately protecting the public in the wake of the Chicago crash of 1979. Nader claims that the FAA has failed to set adequate safety standards, and that there have been failures in the inspection process. He also makes several suggestions for reforms.

Partial confirmation of his criticisms may be found in "Management Improvement Needed in FAA's Airworthiness Directive Program," a Government Accounting Office (GAO) report on weaknesses in the FAA inspection program. It concludes that the FAA's airworthiness directive (AD) program is inadequate since the agency cannot effectively determine whether airlines are complying with airworthiness directives. The report finishes with specific recommendations for improvement.

The excerpt from Charles Perrow's book *Normal Accidents* is the last of the chapters on the FAA. Perrow suggests that the aircraft manufacturers, the airlines, and the FAA support safety modifications and additions only when these lead to increased economic efficiency. This hypothesis is explored and supported in the chapter "FAA, the Carriers, and Safety."

An interesting sidelight on the Chicago crash is that the Airline Passengers Association (now the Airline Passenger Association of

North America) went to court to force the FAA to take action. As a result of their efforts, the DC-10 fleet was grounded—the first such event since 1946—until the problem could be analyzed and solved. This organization and its partner, the International Airline Passengers Association, are strongly critical of the DC-10, and their position is set out in "International Airline Passengers Association Critique of the DC-10," a letter sent to the Transportation, Aviation, and Materials Subcommittee of the U.S. House of Representatives.

In "Moral Responsibility for Engineers," Kenneth Alpern investigates the issue of the moral responsibility of engineers in a more general manner than was done in earlier chapters. He argues that ordinary moral principles impose on engineers the burden of making personal sacrifices for the benefit of society. Practicing engineers, because they exert considerable control over technological developments, can greatly affect public welfare. Since they are in a position to contribute to great harm, it can be expected that they ought to make greater sacrifices than others for the sake of public welfare. The high standard to which we hold engineers is the result of the ordinary requirements of care and proportionate care as they apply to the circumstances of engineers. Alpern does not discuss the DC-10 case in his chapter but it provides another vantage point from which to draw conclusions about the case. It supports our contention that the engineers who design and build planes are a crucial part of the aviation safety system. Andrew Oldenquist argues for a less demanding view of moral responsibility for engineers, one that requires reasonable concessions of self-interest shared nearly equally by all. Samual Florman holds that Alpern's emphasis is misplaced, and points out that engineers do not and should not decide public policy questions concerning safety.

Is the DC-10 an aircraft with an inadequately protected control system, or is it a safe airplane that has had more than its share of bad luck and worse publicity? This is the central ethical issue in the DC-10 case, and how it is judged will strongly influence the ethical evaluation we make of the individuals, organizations, policies, and government agencies that responded to problems with the DC-10. Did these respondents meet their ethical responsibilities at crucial points of the case? Ethical judgments about the participants in the DC-10 case will depend upon whether one views the crucial design decisions, organizational actions, policies, and institutional arrangements as ethically acceptable responses, given the web of engineering, financial, time, organizational, and political constraints that were in place. Ethical decisions and ethical evaluations are strongly context dependent; only by locating a decision or policy within the framework in which it is made can we have an adequate sense of the real options open to the players in the drama.

The chapters in this book will provide the information needed to make ethical judgments about the DC-10 case, but making an ethical evaluation also requires a reflective awareness of the ethical concepts we use in making them. They are the tools we employ to articulate the ethical issues, analyze them and reach ethical decisions. The following section, "Ethical Analysis of Case Studies," provides a introductory discussion of our basic ethical concepts and their application to a case study.

JOHN H. FIELDER

Ethical Analysis of Case Studies

Being ethical has to do with meeting standards of conduct concerning the welfare of others and their right to make their own choices. An ethical person respects others' right to choose, takes their welfare into consideration, and cultivates habits of thought and action (virtues) that promote living an ethical life. Ethical issues in case studies raise questions about how to apply our ethical standards to new or complex situations. They pose difficult questions about right and wrong for which our unreflective ethical norms provide no clear answer. Ethical reflection attempts to articulate ethical standards, critically evaluate them, and apply them to a particular situation, taking into account all ethically relevant circumstances.

The DC-10 case raises many different kinds of ethical questions. The most obvious concerns the choices of various individuals when faced with ethical decisions. But we can also ask whether the actions of organizations, policies, and institutional arrangements reflect an adequate respect for the rights and welfare of persons affected by them.

There is general agreement among ethicists that there are two primary kinds of ethical considerations, those deriving from our status as beings who are able to reason and choose, and others based on our complex physical, social, and emotional needs.[1] The former are emphasized in the ethical tradition of deontology, which focuses on the rights of autonomous human beings to make their own choices without undue interference from others. It is primarily concerned with respect for human choices rather than the good life and emphasizes respect for the rights of others to reason and choose for themselves.[2]

Our welfare has been the central idea of the other ethical tradition, that of consequentialism, which includes the various forms of utilitarianism. Here the emphasis is on how our actions affect the welfare of others by causing benefits and harms. Ethical life in this tradition emphasizes the good and bad effects (consequences) of ethical choices, policies, and institutions.

Many difficult ethical problems stem from conflicts between these two kinds of ethical values. For example, should airline passengers be told that there has been a bomb threat on a particular flight? Doing so would respect their right to choose whether to take the risk of being killed if there is a bomb on board. But there are important negative consequences to the welfare of persons using the air travel system that such a policy of informing passengers would entail. Would terrorists be able to disrupt air travel at will? How many people would be deterred from taking a trip because of frivolous threats? Would the government's intelligence-gathering methods be revealed? Even this brief example shows the complex interplay of individual freedom, different kinds of good and bad consequences, and complex factual conditions that is characteristic of case studies.

These two ethical considerations, human welfare and freedom to choose, are the foundations of our moral tradition. The familiar rules and principles of moral life, such as "Do not lie," are particular expressions of these fundamental ideas. Providing false information by lying interferes with a person's ability to make informed choices and it also may result in loss of goods or opportunities (welfare). Similar analyses can be provided for theft, invasion of privacy, confidentiality, etc.

There has been much philosophical debate about whether one of these ethical traditions is primary and/or implicitly includes the other. A number of ethical theories have been developed in response to this question, but there is no consensus concerning either their adequacy, usefulness for ethical evaluation of case studies, or the validity of the project of ethical theory itself.[3] Consequently, the material in this book does not assume any particular theoretical orientation other than the one outlined in this introduction. Readers who wish to analyze this case using a particular ethical theory are encouraged to do so, but we believe that this is not essential. Familiarity with ethical theory will undoubtedly bring additional insight into some aspects of the case, but so would a knowledge of law, organizational behavior, economics, political science, or the history of aviation.

CASE STUDIES

The ethical analysis of case studies is concerned with making ethical judgments about actions, choices, policies, and institutions in actual situations. The focus is on ethical evaluation that takes into account the complexity of all of the ethically relevant circumstances. The analysis of case studies has many similarities to the legal tradition of common law, which is based on the interpretation of precedent cases rather than statutes. Those pursuing a career in law use both actual and hypothetical case studies in learning to apply legal standards of conduct to unusual, unforeseen, and complicating circumstances of real life. Case analysis is often used in other professional programs, such as the Harvard Business School, because of its value in preparing students to deal with the kind of problems they will encounter in actual practice. It provides an opportunity to learn to apply management concepts and principles to the complex problems of business managers. Ethicists similarly apply ethical principles and concepts to actual situations, taking account of the ethically relevant circumstances.

The DC-10 case invites ethical evaluation on a wide range of levels. Did McDonnell Douglas design an adequately safe airplane? Were McDonnell Douglas's actions in responding to problems with the DC-10 ethically acceptable? Should Dan Applegate have blown the whistle? Was it wrong for the FAA to have made a deal with McDonnell Douglas after the Windsor incident? Does the policy of using industry employees to make FAA inspections meet our ethical standards? Is it right to have an institutional arrangement which assigns the FAA the role of both promoting the aviation industry and ensuring the safety of air travel? Is capitalism with government regulation adequate to ensure the level of safety that air passengers have a right to expect? These are the kinds of questions the DC-10 case raises. What kinds of conceptual tools can philosophy provide to answer them?

It is important to keep in mind that these questions can only be posed because we already have an ethical vocabulary in which we can ask ethical questions, propose answers, and debate them. We begin our ethical analysis already in possession of a set of shared ethical principles and concepts and experience in using them. All of us start from the belief that it is generally wrong to lie, cheat, steal, kill, invade people's privacy, inflict physical or emotional harm on them, prevent them from making their own life decisions, etc. We have had extensive experience in using these ideas to analyze ethical problems, make our own ethical judgments and criticize the judgments of others. Making ethical judgments about persons, actions, policies, and institutions is an everyday feature of life.

While philosophers disagree about the role of ethical theory in analyzing case studies, they share a universal belief that ethical analysis can be strengthened by clarifying the important concepts we use in ethical analysis. This helps to eliminate confusion, allows us to be more explicit about our analysis, and more accurately frame important points of disagreement. The remainder of this chapter sets out some of our basic ethical concepts and outlines their connections to the DC-10 case.

ETHICAL ANALYSIS: OBLIGATIONS AND RIGHTS

Ethical analysis of actual ethical problems is primarily a matter of articulating standards of ethical conduct and applying them to the particular circumstances of the case. It is essential to articulate the ethical requirements relevant for the evaluation of a particular individual or organizational action, policy, or institution. Two concepts that are of central importance for this task are *obligations* and *rights*.

Obligation is a concept that occurs in law, ethics, religion, and social life. It designates requirements of conduct that arise within a particular tradition or practice. Legal obligations arise from the framework of laws that govern our actions, and in a similar way religious traditions have specific requirements of conduct for their adherents. Membership in a community also creates social obligations to neighbors and others with whom we interact. Ethical obligations arise from our participation in a community with shared norms of behavior. Whether we view those norms as having transcultural validity (as traditional ethical theories hold) or as the evolved values of Western civilization (as some of the critics of ethical theory believe), these norms may be stated as obligations requiring certain kinds of conduct.

An obligation may usually be restated in terms of a right. For example, a teacher's obligation to protect the privacy of students' grades may be equivalently stated as the students' right to have the privacy of their grades protected. The ethical requirement can be stated from the point of view of those who act (their obligations) or those who are importantly affected by those acts (their rights). Respect for the rights of others means meeting the obligations those rights impose upon us.

The language of rights is most closely associated with the deontological tradition of ethics, which emphasizes freedom of choice. Rights in this usage assert special protections against the interference of others. But this concept is now applied more broadly, so that it may also refer to claims others have on us concerning their welfare.

POSITIVE AND NEGATIVE RIGHTS AND OBLIGATIONS

Most of the obligations we learned while growing up were negative, that is, they stipulated that one was not to do certain things: lie, cheat, steal, harm others. They were "don'ts." These obligations can be stated as negative rights that require others not to interfere with us in certain ways, such as taking our possessions, causing us pain, or causing us to believe something false. These rights are primarily designed to protect us from the actions of others. It is the idea behind the familiar ethical principle that "your right to swing your fist ends where my nose begins."

While negative rights and obligations require that we refrain from certain actions, positive rights and obligations require that we perform certain actions. For example, parents are obligated to do many things to ensure that their children's needs are met. Proper diet, schooling, love and nurturing are positive actions that parents owe their children. These obligations may be restated as positive rights, which are sometimes called welfare rights. Children have a right to an adequate diet, and it is not only parents who have corresponding obligations to see that they get it. Social and governmental institutions can powerfully affect the lives of children by making it easy or difficult for parents and others to meet their obligations to them. Organizations and institutions as well as individuals may be held ethically accountable for their roles in the ethical treatment of children.

The importance of this concept in aircraft safety concerns the welfare rights of passengers. If they have a positive welfare right to safe air travel, then all those who have a role in providing air travel have corresponding positive obligations to see that air travel is safe. This will include not only individuals but also corporations, policies, government agencies, and fundamental institutional arrangements. All must meet ethical obligations if the passenger has such a right. Much social criticism consists of showing that generally accepted positive rights to safety, health care, education, etc. are not being respected by our current social arrangements.

ENGINEERING AS SOCIAL EXPERIMENTATION

A useful way to assert the ethical rights of passengers in the DC-10 case is to use the idea of engineering as a form of social experimentation.[4] Since it is impossible to foresee all of the kinds of accidents that can occur in a sophisticated engineering project like a new aircraft, putting it into commercial service is a kind of social experiment. It is expected that unforeseen problems will emerge and that there will be modifications as a result of "testing" the aircraft by putting it into service and flying it under all kinds of conditions. C. O. Miller, former director of aviation safety at the National Transportation Safety Board, put it this way: "The operating world has a unique capacity to make mistakes that no designer ever thought could be made. The complexity of modern technology is such that I defy anybody—anybody, no matter how good he is—to predict all operating errors."[5]

The passengers and flight crew are, in a real sense, subjects taking part in the experiment. The testing of a new drug, for example, takes place only after numerous animal and other studies have revealed much about the drug's properties, but its effects on humans is not entirely predictable. Only tests on human beings will determine how it will actually affect people's health. By participating in an experiment, the subjects take on a risk that unforeseen difficulties can be harmful.

Aircraft are extensively analyzed and tested before being sold to airlines, but there are also many unknowns about their performance under a wide range of commercial and physical conditions. One of the more dramatic examples of this concerned the de Havilland Comet jet aircraft of the 1950s. Unforeseen metal fatigue caused the wing to separate from the fuselage after several months of operation.[6] The introduction of this and other kinds of jet aircraft were social experiments that had many unexpected outcomes, some of which were fatal to the passengers.

Viewing air travel as a social experiment means that the participants in that experiment—passengers and flight crews—are entitled to have adequate safety provisions made in the design and use of the aircraft. This, in turn, means that individuals, organizations, policies, and institutions have positive ethical obligations to respect the rights of air travelers concerning aircraft safety.

SAFETY

Safety can be defined as "of acceptable risk."[7] Risk is a combination of the probability of harm and the severity of harm. This captures our intuition that the most serious risks are those that involve both severe harm and a high probability that it will happen. This aspect of risk analysis can be carried out by experts who are trained to estimate the probability of occurrence and assess the kind of harms that would result.[8]

Whether or not something is safe depends upon whether the risk is "acceptable," and this is clearly not a technical question to be answered by experts. It is a social and political question in its application to air travel, for it is ultimately the public that should determine what kinds of risks are acceptable in boarding aircraft. In the case of air travel the public's agents—the FAA and the other agencies—make decisions about safety, but they are certainly open to challenge. Many critics of the FAA believe that aviation is not sufficiently safe, that the public has a right to more protection than has been provided in the case of the DC-10.

PRESUMPTIVE RIGHTS AND OBLIGATIONS

Rights and obligations are generally regarded as *presumptive*, meaning that they are assumed to be binding but may be set aside under certain conditions or "trumped" by more important rights or obligations. This is the familiar idea that ethical principles have exceptions. For example, although lying is assumed to be wrong, it is not difficult to imagine conditions under which it would not be. Lying to the Gestapo to protect Jews is morally right because the obligation to protect innocent lives is more important in that situation than the obligation to tell the truth. Both are obligations, but there seems to be no response which will fulfill both of them. When there is a conflict of obligations and there is no way to meet both, we have to decide which obligation should have priority. Many ethical disagreements concern issues where each side has a valid moral claim and the conflict concerns which is more important.

JUSTIFICATIONS, EXCUSES, AND SACRIFICE

Persons who defend their violation of the obligation to tell the truth (as in the example of lying to the Gestapo to save Jews), argue that their actions are *justified* because they were not simply lying but were meeting a more important obligation, that of preserving innocent life. Since there was no way to meet both obligations, their violation of the less important one was morally justified.

A violation of a moral principle may also be *excused*. Suppose a friend promised to meet you for lunch but on the way her car broke down. She has good reason to claim that the failure to keep the promise is excused because of circumstances beyond her control. Of course, she could have abandoned the car, hailed a cab and kept the promise, but we would not require this amount of sacrifice of her other personal goals (e.g., to get the car fixed as quickly as possible, avoid extra expense, etc.) to keep this kind of obligation. But what if she was maid of honor at your wedding? Or the minister? Under either of these circumstances we would expect her to make more of an effort to attend because of the greater importance of the obligation. Those efforts would require that she devote time, effort, and money to meeting her obligation that could otherwise have been devoted to other activities. The greater the importance of the obligation the greater the amount of sacrifice or risk taking it requires. The importance of an obligation and the amount of risk or loss that would result from meeting it are key factors in determining what counts as an adequate excuse.

Although Dan Applegate warned his superiors that the DC-10's cargo door was dangerous, they took no action and he apparently did not pursue the matter. Assuming that Applegate had an obligation to the passengers of the DC-10 to protect them from this kind of danger, were his actions sufficient to meet his obligation? If taking further action would cause substantial danger to his career, does that danger excuse his failure to press the issue and possibly blow the whistle? In other words, how much sacrifice of one's personal goals can we ask someone in Applegate's position to make?

ROLE RESPONSIBILITIES

The most obvious place where ethical obligations apply is individual decisions. In a particular ethical problem we are not so much concerned with broad ethical principles (such as "Do not lie") that apply to all individuals but with how those principles are applied in particular circumstances. An important feature of the circumstances in which ethical problems arise is the *role* a particular person has. The concept of *role responsibility* helps to clarify the obligations people have by deriving some of them from their roles.[9]

According to this approach, each role embodies a set of rights and obligations—ethical, social, and legal—in a social or organizational function. For example, a lifeguard has a greater ethical obligation to help persons who may be drowning than the vacationers walking along the beach. Similarly the lifeguard's role entails legal obligations concerning the upkeep of lifesaving equipment which do not apply to others, and social requirements that arise in the community of lifeguards and persons associated with that role. Determining whether individuals acted ethically is often a matter of deciding what the person's relevant role obligations are, whether or not they were fulfilled, and whether there were any excusing or justifying conditions present.

After a DC-10 lost its cargo door over Windsor, Ontario, the FAA administrator (the head of the agency) chose to make an informal deal with McDonnell Douglas rather than issue an airworthiness directive. Did this action meet the ethical role responsibilities of a person in that situation? This question is complicated by the fact that the FAA's charter requires its administrator to both promote aviation and assure its safety. Thus the administrator has potentially conflicting role responsibilities. This example shows how policies and institutional arrangements create role responsibilities that place great, if not impossible, ethical demands on the persons who fill them.

PROFESSIONAL OBLIGATIONS

An important subset of role obligations are *professional obligations*. These are obligations that apply to special kinds of careers which are called professions. Law, medicine, nursing, and engineering are typical examples of professions. They are characterized by (a) the possession of specialized skills and knowledge which is (b) obtained through extensive, formal education involving substantial theoretical content, and results in (c) a service performed by the practitioner that is essential to the community.[10] Because professionals possess essential specialized skills and knowledge not available to lay persons, the latter are relatively ignorant and dependent upon professionals for the services based on their specialized knowledge. Because of this unequal relationship, special ethical obligations apply to professionals which are designed to protect their vulnerable clients and to ensure that the policies and institutions associated with professional services are also ethically acceptable. In the DC-10 case, professional engineers had knowledge about the deficiencies of the aircraft which lay persons did not possess. Because that knowledge was of potential harm to the innocent (and ignorant) passengers and flight crew, the engineers have a professional obligation to take action to try to remove the harm.

The profession's understanding of those obligations is expressed in codes of ethics. Engineering codes of ethics have addressed the problem of engineers like Dan Applegate, who know about danger to others. For example, the Institute of Electrical and Electronic Engineering (IEEE) Code requires that "Members shall, in fulfilling their responsibilities to the community: protect the safety, health and welfare of the public and speak out against abuses in these areas affecting the public interest" (Article IV, 1; this code of ethics is included in the Appendix). Other codes are more forceful about the priority of this obligation, stating that the obligation to public safety, health, and welfare is "paramount." These codes of ethics clearly require engineers to take action to protect innocent third parties from danger, but they leave it to the individual's judgment to decide what action is appropriate.

ORGANIZATIONAL ACTIONS

Ethical standards apply not only to individual decisions but also to organizational actions, to policies, and to institutional arrangements. Organizational actions, unlike individual ones, embody decisions made by groups of people acting within organizational roles and exercising organizational authority. These organizational decisions affect the rights and welfare of others. The ethical questions that arise here will concern whether organizational actions adequately embody respect for the rights and welfare of those who will be affected by them. As with individual decisions, much depends upon the specific character of organizational actions, their overall effects, the options available, and the constraints present.

Convair officials read Applegate's memo, but the organization took no action. Was this an ethically acceptable response? Should Convair have taken further action to try to get the cargo door problem resolved? Similar questions can be asked about various actions taken by McDonnell Douglas and the FAA when they were warned about the dangers associated with the DC-10.

Organizational actions differ from individual choices in that the former are typically the result of interactions among individuals whose position is defined by their roles in the organization. The resulting action reflects both individual choices and the organization's history, culture, and structure.[11]

POLICIES

Policies are larger patterns of organizational action. They can be evaluated according to whether or not they meet the obligations of the organization to those who will be affected by the policy. A policy may be explicitly spelled out in memos or other forms of organizational publications, such as personnel manuals, or it may be implicit, reflecting "how things are done around here" without being formally set out. Both represent organizational decisions about how to deal with a certain kind of issue and are subject to ethical evaluation.

The FAA has a policy of having most of the inspections of aircraft performed by employees of the manufacturer. These designated engineering representatives (DERs) serve both the regulating agency (the FAA) and the regulated industry (McDonnell Douglas). Many observers, including Ralph Nader, have pointed out the clear conflict of interest this policy embodies and have questioned its ethical acceptability.

Ethical objections may also be made to the FAA's policies concerning airworthiness directives (ADs). As the Government Accounting Office (GAO) report shows, the FAA does not have an adequate policy of inspection to ensure that the ADs are in fact carried out as ordered. Inspectors have more items that are subject to inspection than time allows, so they can use their own judgment about what to inspect and what to ignore. The GAO report challenges the adequacy of that policy by pointing out examples of ADs that were not carried out and posed unnecessary dangers to the passengers.

Policies may be explicit, as in these examples, or they may be unstated patterns of action. McDonnell Douglas consistently sought to minimize and delay design changes on the DC-10. It is not likely that this was an explicit policy, but it clearly functioned in the same way, guiding organizational actions. In the eyes of some critics, this kind of unstated policy is more ethically questionable than specific actions that followed from it.

INSTITUTIONAL ARRANGEMENTS

We may also direct ethical evaluation to the fundamental political arrangements in which the aviation industry functions. In the United States the aircraft manufacturers and suppliers are private businesses operating within a form of a capitalist economic framework. Making sure that the airplanes and their operation by the airlines is adequately safe is primarily the responsibility of the FAA. Critics like Ralph Nader and Charles Perrow have questioned the ability of this structure to meet the obligations to the consumers of air travel. The FAA's charter requires it to both promote the aircraft industry and to ensure the safety of its products. This feature of our institutional arrangements has been criticized as inadequately protecting the flying public by allowing safety to be compromised if the economic health of the industry is threatened. Some see this problem as needing a restructuring of institutional relationships, such as transferring responsibility for promotion of the aviation industry to another division of the government. Those who believe that the capitalist system itself is the problem will provide more radical proposals.

CONCLUSION

The ethical analysis of case studies demands both clarity about ethical concepts and familiarity with the actual circumstances in which we apply them. The DC-10 case is extremely complex and ethically interesting, involving high technology, government bureaucracies, powerful corporations and the individuals who work for them, various policies which attempt to meet sometimes conflicting values, and our basic institutional arrangements for ensuring the safety of air travel. Because it touches so many different dimensions of contemporary commercial life, it generates a variety of difficult ethical questions. The ethical concepts, the information, and the analysis of ethical issues collected in this book will not answer all of them in a definitive way, but they will enable you to develop your own well-informed judgments about a difficult problem in contemporary life.

NOTES

1. See Louis Lombardi, *Moral Analysis* (Albany, NY: SUNY Press, 1988) for an extended treatment of this issue.

2. See Jeffrey Stout, *Ethics After Babel* (Boston: Beacon Press, 1988), p. 286.

3. The debate about the role of ethical theory in philosophy is extensive. Good places to enter that literature are: *Anti-Theory in Ethics and Moral Conservatism*, edited by Stanley G. Clarke and Evan Simpson (Albany, NY: SUNY Press, 1989); *Morality and Conflict*, by Stuart Hampshire (Cambridge: Harvard University Press, 1983); *The Abuse of Casuistry*, by Albert Jonsen and Stephen Toulmin (Berkeley: University of California Press, 1988).

4. This idea is developed in *Ethics in Engineering*, by Mike W. Martin and Roland Schinzinger (New York: McGraw-Hill, 1983), p. 55 ff.

5. J. M. Ramsden, *The Safe Airline* (London: MacDonald and Jane's, 1976), p. 25.

6. See *Great Air Disasters*, by Stanley Stewart (London: Arrow Books, 1988), pp. 54-90.

7. William W. Lowrance, *Of Acceptable Risk* (Los Altos, CA: William Kaufman, Inc., 1976), p. 8.

8. The public perception of risk is not always the same as that of the experts. See Paul Slovic, Baruch Fishchoff, and Sara Lichtenstein, "Facts and Fears: Understanding Perceived Risk," in *Societal Risk Assessment: How Safe is Safe Enough?*, edited by Richard C. Schwing and Walter A. Albers (New York: Plenum Press, 1980).

9. See Norman F. Bowie and Ronald F. Duska, *Business Ethics*, 2nd Edition (Englewood Cliffs, NJ: Prentice-Hall, 1990), p. 4 ff. This idea is also developed in Louis Lombardi, *Moral Analysis*.

10. For a discussion of this issue as it applies to engineers, see D. Allan Firmage, *Modern Engineering Practice: Ethical, Professional, and Legal Aspects* (New York: Garland STPM Press, 1980), pp. 10-14.

11. For a contrary view which holds that it makes no sense to apply concepts of moral responsibility to organizations, see Manuel G. Velasquez, "Why Corporations are Not Morally Responsible for Anything They Do," *Business and Professional Ethics Journal*, 2 (Spring 1983): 1-17.

HISTORY AND EARLY WARNINGS

1

Regulatory and Institutional Framework*

The Federal Aviation Administration (FAA) has a dual mandate: ". . . to promote safety of flight . . . in air commerce through standard setting . . ." and to encourage and foster the development of air commerce.[1] The Airline Deregulation Act, passed in 1978 to encourage industry competition, removed federal controls over routes, fares, and new entries, but left unaltered the FAA's responsibility for commercial aviation safety. Events of the past decade have shown that neither Congress nor the executive branch fully comprehended the complexity of regulating a newly competitive industry. Although commercial aviation maintains an enviable safety record, dramatic growth in air travel, major changes in technology and industry operations and structure, the firing of the air traffic controllers, and federal budget constraints have left FAA

---

* Reprinted with editorial changes, from *Safe Skies for Tomorrow: Aviation Safety in a Competitive Environment,* U.S. Congress, Office of Technology Assessment (Washington, D.C.: U.S. Government Printing Office, 1988).

29

scrambling to catch up. Consequently, public attention has again focused sharply on whether FAA has the institutional capability and resources to carry out its operating, standard setting, rule making, and technology development functions effectively and to guarantee compliance through its inspection programs.

Before 1978, the relative stability of the commercial airline industry made carrying out FAA's regulatory activities less contentious. Industry changes occurred slowly, fewer carriers were competing for the travel dollar, and the costs of required safety improvements could be passed quickly to the consumer. Today's environment is dramatically different, forcing FAA to oversee an industry in which major players come and go, and airlines must expand markets and control labor and other operating costs carefully or go bankrupt. One consequence is that aircraft manufacturers and airlines scrutinize critically any changes of safety regulations, especially those requiring expensive new technology or additional personnel training. Moreover, federal policies have explicitly discouraged new regulation, unless judged cost-effective, while local government policies have restrained new airport development. This chapter provides an overview of the evolution of federal aviation safety laws and regulations, describes the current institutional framework, provides analyses of the FAA safety programs, and the impact of local regulations on airport use and development.

HOW IT ALL BEGAN

The roots of today's aviation safety programs, including their rough edges, extend back to the early days of aviation in the mid-1920s. Early commercial uses of aircraft included advertising, aerial photography, crop dusting, and carrying illegal shipments of liquor during Prohibition. Initial efforts to establish scheduled passenger service were shortlived, as service catered primarily to wealthy east coast tourists and was expensive relative to the country's well-developed rail and water travel networks.

Air Mail Service

Growth of commercial aviation was greatly stimulated by the establishment of the U.S. Air Mail Service in the early 1920s. Regulations established by the Post Office Department required its pilots to be tested and to have at least 500 hours of flying experience and set up aircraft inspection and preventive maintenance programs. These early regulatory requirements improved air mail carrier safety—in 1924, commercial flyers experienced one fatality per 13,500 miles, while the Air Mail Service had one fatality per 463,000 miles.[2]

In 1925, Congress enacted the Air Mail Act, authorizing the Post Office Department to transfer air mail service to private operators. Twelve carriers, some of which evolved into today's major airlines, began air mail operations in 1926 and 1927. These carriers offered limited passenger service, which was much less profitable than carrying mail.[3] Small independent operators, using Ford and Fokker tri-motor airplanes, handled most of the passenger service in the late 1920s, the forerunners of today's commuter airlines and air taxis.

Early Safety Initiatives

No federal safety program existed, prompting a number of states to pass legislation requiring aircraft licensing and registration. In addition, local governments of all sizes enacted ordinances regulating flight operations and pilots, creating a patchwork of safety-related requirements and layers of authority. Modern versions of these difficulties are discussed later in this chapter. Despite strong industry support for federal legislation, Congress was unable to reach agreement on the scope and substance of a statute until 1926,[4] when the Air Commerce Act was passed.[5] The new law charged the Department of Commerce with both regulatory authority over commercial aviation and responsibilities aimed at promoting the fledgling industry. The major provisions of the act authorized the regulation of aircraft and airmen in interstate and foreign commerce; provided federal support for charting and lighting airways, maintaining emergency fields, and making weather information available to pilots; authorized aeronautical research and development programs; and provided for the investigation of aviation accidents. Local governments were left with jurisdiction over airport control.

Within the Department of Commerce, a new Aeronautics Branch, comprised of existing offices already engaged in aviation activities, was formed to oversee the implementation of the new law. Nine district offices of the Regulatory Division of the Aeronautics Branch were established to conduct inspections and checks of aircraft, pilots, mechanics, and facilities, and share licensing and certification responsibilities with the Washington, D.C. office. The basic allocation of responsibilities survives to this day, although the Department of Commerce responsibilities now rest with the Department of Transportation (DOT) and its arm, FAA.

The first set of regulations was drafted with substantial input from aircraft manufacturers, air transport operators, and the insurance industry. Compared with current standards, pilot requirements were minimal; in addition to written and flight tests, transport pilots were required to have one hundred hours of solo flight experience, while industrial pilots needed only fifty hours.

Current procedures for certifying aircraft and engines also originated under these early regulatory programs. Aircraft manufacturers were required to comply with minimum engineering standards issued by the Department of Commerce in 1927, and one aircraft of each type was subject to flight testing to obtain an airworthiness certificate for the type.

The Aeronautics Branch also collected and analyzed data from aircraft inspection reports, pilot records, and accident investigations. These data were made accessible to the insurance industry, allowing the development of actuarial statistics. A direct consequence of this step was a significant reduction in insurance rates for many carriers. However, the Department of Commerce, cognizant of its role to promote the aviation industry, was reluctant to make public disclosures about the results of individual accident investigations, despite a provision in the 1926 act directing it to do so. Eventually, in 1934, the Air Commerce Act was amended, giving the secretary of commerce extensive powers to investigate accidents, including a mandate to issue public reports of its findings.[6] This congressional policy decision put safety considerations ahead of protecting the industry's image.

As additional regulations to improve safety were implemented, accidents involving passenger carriers and private aircraft decreased significantly; between 1930 and 1932, the fatality rate per 100 million passenger-miles declined by 50 percent.[7] Updated regulations established more stringent requirements for pilots flying aircraft in scheduled interstate passenger service, including flight-time limitations.[8] Other requirements specified the composition of flight crews, established standards for flight schools, improved takeoff and landing procedures, set minimum flight altitudes and weather restrictions, and required multi-engine aircraft to be capable of flying with one inoperative engine. In addition, certification of carriers providing scheduled passenger service in interstate commerce commenced in 1930. Although financial data were not examined by the Department of Commerce, standards for key personnel, the ground organization of a carrier, maintenance procedures, and aircraft equipment and instruments had to be met.[9]

The Beginning of Economic Regulation

During the 1930s, industry expansion and the development of aircraft and communication technologies required continuous improvements of regulations, airways, and airports. However, budget constraints prevented the Department of Commerce from conducting sufficient inspections and keeping up with airway development needs. Moreover, a series of fatal accidents in late 1935, 1936, and 1937, including one in New Mexico that killed a New Mexico senator, called into question the adequacy of existing regulations.[10]

The Civil Aeronautics Act of 1938 marked the beginning of economic regulation. It required airlines, with or without mail contracts, to obtain certificates authorizing service on specified routes, if the routes passed a test of public convenience and necessity.[11]

The act created the Civil Aeronautics Authority (CAA), which was responsible for safety programs and economic regulations, including route certificates, airline tariffs, and air mail rates. Within CAA, a separate Administrator's Office, answering directly to the president, was responsible for civil airways, navigation facilities, and controlling air traffic.[12] However, in June 1940, under the Reorganization Act of 1939, CAA was transferred back to the Department of Commerce and the Civil Aeronautics Board (CAB) was created and made responsible for regulatory and investigatory matters.

An Expanding Federal Role

Federal responsibilities for airway and airport development grew tremendously during World War II, leading to passage of the Federal Airport Act of 1946, and initiating federal financial assistance to states and municipalities. The federal government assumed responsibility for air traffic control (ATC) at this time. However, the inspector force could not keep pace with the rapidly increasing numbers of new airplanes, pilots, and aviation-related facilities. As early as 1940, CAA had designated certain parts of the certification process to industry. For example, flight instructors were permitted to certificate pilots, and a certificated airplane repaired by an approved mechanic could fly for thirty days until it was checked by an available CAA inspector. After the war, CAA limited its aircraft certification and inspection role to planes, engines, and propellers; manufacturers became responsible for ensuring that other aircraft parts met CAA standards.[13]

Decentralized Management. Regulatory and organizational changes also took place during and after the war. Regional offices of

CAA, reduced in number to seven in 1938, became more autonomous in 1945. Regional officials became directly responsible for operations in their regions, although technical standards and policies were still developed in Washington, D.C. Except for a brief return to more centralized management in the late 1950s, regional autonomy within FAA has persisted to this day, slowing communications between and among headquarters and the regions and intensifying inequities in regulatory applications.

Updating Regulations. Fatal crashes in the late 1940s and early 1950s prompted revised standards setting minimum acceptable performance requirements, designed to ensure continued safe flight and landing in the event of failure of key aircraft components. These standards also distinguished small and large airplanes based on existing airplane and powerplant design considerations; small airplanes were those with a maximum certificated takeoff weight of 12,500 pounds or less, while airplanes above 12,500 pounds were defined as large.[14] This distinction is still applied by FAA today, despite significant changes in aircraft design.

Industry Expansion

Beginning of Air Taxi Service. Surplus war transport airplanes and a new supply of pilots led to the development of the nonscheduled operator or air taxi. Exempt from economic regulation by the Civil-Aeronautics Act of 1938, these operators transported persons or property over short distances in small airplanes, often to locations not serviced by the certificated airlines. CAA, at the time sympathetic to private and small operators, applied less stringent safety regulations to air taxis.[15] In 1952, exemption from economic regulation became permanent, even for carriers using small aircraft to provide scheduled service.[16]

Certificated Airlines. The decade following World War II witnessed enormous industry growth. Pressurized aircraft traveling at greater speeds and carrying more passengers were introduced.[17] In addition to scheduled passenger service, air freight operations expanded when CAB granted temporary certificates of public convenience and necessity to four all-cargo airlines in 1949.[18] Certification and operating rules for commercial operators—those offering contract air service for compensation or hire—were also adopted in 1949.[19]

Responding to Industry Growth

However, despite continuing increases in air traffic and the need for better airports to accommodate larger and faster aircraft, federal support for ATC facilities, airport development, and airway modernization was insufficient. CAA, faced with budget reductions in the early 1950s, was forced to abandon control towers in eighteen small cities and numerous communications facilities, postpone jet development and navigation improvements, and curtail research efforts. The federal airport development program, championed by cities and smaller municipalities, was embroiled in controversy. In addition, the number of CAA regional offices was reduced from seven to four, thirteen safety inspection field offices were eliminated, and the industry designee program was expanded.

The impending introduction of jet aircraft and a 1956 midair collision over the Grand Canyon involving a DC-7 and a Super Constellation helped promote congressional authorization of increased levels of safety-related research and more federal inspectors. In 1958, Congress passed the Federal Aviation Act establishing a new aviation organization, the Federal Aviation Agency.[20] Assuming many of the duties and functions of CAA and CAB, the agency was made responsible for fostering air commerce, regulating safety, all future ATC and navigation systems, and airspace allocation and policy. CAB was continued as a separate agency responsible for economic regulation and accident investigations.[21]

The safety provisions of the 1958 act, restating earlier aviation statutes, empowers the agency to promote flight safety of civil aircraft in air commerce by prescribing:[22]

- minimum standards for the design, materials, workmanship, construction, and performance of aircraft, aircraft engines, propellers, and appliances;
- reasonable rules and regulations and minimum standards for inspections, servicing, and overhauls of aircraft, aircraft engines, propellers, and appliances, including equipment and facilities used for such activities. The agency was also authorized to specify the timing and manner of inspections, servicing, and overhauls and to allow qualified private persons to conduct examinations and make reports in lieu of agency officers and employees;
- reasonable rules and regulations governing the reserve supply of aircraft, aircraft engines, propellers, appliances, and aircraft fuel and oil, including fuel and oil supplies carried in flight;

- reasonable rules and regulations for maximum hours or periods of service of airmen and other employees of air carriers; and
- other reasonable rules, regulations, or minimum standards governing other practices, methods, and procedures necessary to provide adequately for national security and safety of air commerce.

In addition, the act explicitly provides for certification of airmen, aircraft, air carriers, air navigation facilities, flying schools, maintenance and repair facilities, and airports.[23]

In the years following creation of the agency, federal safety regulations governing training and equipment were strengthened despite intense opposition from industry organizations. The number of staff members also grew in the early 1960s, and inspection activities were stepped up, including en route pilot checks and reviews of carrier maintenance operations and organizations.[24]

In 1966, the Federal Aviation Agency became the Federal Aviation Administration, when it was transferred to the newly formed Department of Transportation (DOT).[25] The National Transportation Safety Board (NTSB) was also established to determine and report the cause of transportation accidents and conduct special studies related to safety and accident prevention; accident investigation responsibilities of CAB were moved to NTSB.

Renewed support for improvements to airports, ATC, and navigation systems was also provided by the Airport and Airway Development Act of 1970. The act established the Airport and Airway Trust Fund, financed in part by taxes imposed on airline tickets and aviation fuel, and was reauthorized in 1987.[26]

Recognizing that existing industry descriptors, such as trunks, locals, and commuters were no longer appropriate, CAB redesignated scheduled passenger airlines into the following groups based on annual revenues:

- major airlines (above $1 billion);
- national airlines ($75 million to $1 billion);
- large regional airlines ($10 million to $75 million); and
- medium regional airlines (up to $10 million).[27]

Airline Deregulation

Prompted by widespread dissatisfaction with CAB policies and the belief that increased competition would enhance passenger service and reduce commercial airline fares, Congress enacted the Airline Deregulation Act of 1978.[28] Specifically, the act phased out over a six-year period CAB control over carrier entry and exit, routes, and fares. In 1984, the remaining functions of CAB were transferred to DOT. These functions include performing carrier fitness evaluations and issuing operating certificates, collecting and disseminating financial data on carriers, and providing consumer protection against unfair and deceptive practices.[29]

During the sixty-year history of federal oversight, federal regulatory and safety surveillance functions have been frequently reorganized and redefined. Moreover, public concerns about how FAA carries out its basic functions have remained remarkably constant despite a steadily improving aviation safety record. The Office of Technology Assessment's brief historical summary demonstrates that:

- Specifically authorized by legislation after lengthy debate in the 1950s, industry participation in regulatory activities has a long history. Responsible federal aviation agencies consistently have designated part of the certification and inspection processes to the private sector, specifically certification of pilots, aircraft parts, and aircraft repair. This reliance on private industry is heaviest when national budget constraints lead to shortages of federal inspectors and technical expertise.
- From the initial 1926 legislation to the 1978 Deregulation Act, Congress has charged federal agencies with the dual responsibilities of maintaining aviation safety and promoting the industry, which history shows are not always compatible. Furthermore, except for a two-year period from 1938 to 1940, federal regulatory and enforcement functions have been combined in one agency.
- Federal aviation regulatory enforcement activities have always been decentralized with regional and district offices having considerable autonomy and independence from Washington headquarters.
- More stringent safety standards usually follow a widely publicized airline accident and vocal public and congressional concern, rather than from FAA initiatives. Recent examples include legislation enacted in late 1987 of collision avoidance equipment requirements for commercial aircraft and broadened altitude encoding transponder requirements for general aviation aircraft.

FEDERAL AVIATION SAFETY STRUCTURE

FAA Responsibility

Since Congress dismantled CAB, FAA has been the chief regulator of the U.S. airline industry, with some political and analytic support from other parts of DOT. The task is formidable. On the one hand, the agency must stand up to intense pressure from DOT and industry on proposed regulatory changes, and, on the other, address constant public and congressional anxieties about safety and convenience. FAA's effectiveness has been undercut by budget constraints affecting personnel and procurement, equipment obsolescence, inadequate, long-range, comprehensive planning, and problems with its inspection and rule-making programs. (Table 1-1 shows the impacts of budget constraints on personnel levels in critical areas.) Furthermore, local governments play major roles in determining airport operations and development, often conflicting with FAA goals. Only an agency with strong leadership and singleness of purpose and responsibility could maintain a steady course under such conflicting pressures.

Although all FAA sections have safety-related activities, responsibility for the largest safety programs is under the purviews of the associate administrators for Air Traffic, Aviation Standards, and Development and Logistics. Also, all nine regional offices have broad and separate authority, as does the Mike Monroney Aeronautical Center in Oklahoma City. This splintering of authority has long been recognized as creating fundamental organizational problems within FAA and in its relationship to Congress, DOT, and industry.

- Aviation Standards. Headquartered in Washington, Aviation Standards manages field offices in charge of both airworthiness standards for aircraft and regulations for all air carrier operations. The Aviation Standards National Field Office, located in Oklahoma City, has responsibility for a variety of support activities, including management of national safety databases and conduct of standardization training for designated examiners. Aviation Standards also receives technical support from the FAA Technical Center in Atlantic City, New Jersey, for regulatory development and for research and testing related to crashworthiness and fire safety.
- Air Traffic. Through the regions, Air Traffic is responsible for

TABLE 1.1
Selected FAA Employee Totals, 1978 to 1987

Occupation	1978	1979	1980	1981	1982	1983	1984	1985	1986	1987
Air traffic controller[a]	16,750	16,853	16,584	6,658	11,416	11,946	11,944	12,245	12,429	12,847
Aviation safety inspector[b]	1,466	N/A	1,499	1,615	1,423	1,331	1,394	1,475	1,813	1,939
Electronics technician[c]	9,423	9,209	8,871	8,432	8,031	7,633	7,229	6,856	6,600	6,740

[a] Full performance level and developmental controllers at towers and centers.
[b] Air carrier inspectors (approximately 40 percent of the total) were responsible for 145 air carriers, while general aviation inspectors were responsible for 173 Part 135 commuter airlines, 7,804 other commercial aircraft operators, and 5,210 aviation schools and repair stations as of March 10, 1988.
[c] Communications, navigational aid, radar, and automation technicians.

SOURCE: Office of Technology Assessment based on Federal Aviation Administration data as follows: controller data as of September 1987; inspector data as of March 1988; and technician data as of March 1988.

operation of the 20 Air Route Traffic Control Centers, 176 Terminal Radar Approach Control facilities, hundreds of airport towers, the Central Flow Control Facility, plus flight service stations located throughout the United States and Puerto Rico. In addition, Air Traffic formulates plans and requirements for future ATC operations, and evaluates and analyzes current ATC operations.

- Development and Logistics. Development and Logistics is in charge of technology development, implementation, and maintenance, and has overall responsibility for the National Airspace System (NAS) Plan. Offices within Development and Logistics include Automation Service, which is in charge of upgrading the ATC system and implementing the Advanced Automation System. Program Engineering Service directs other NAS Plan programs, and Systems Engineering Service handles system engineering for the NAS Plan, advanced systems and concepts, and development of the NAS Performance Analysis Capability for systemwide airspace management. Systems Maintenance Service directs maintenance of the NAS. The FAA Technical Center performs engineering and testing for NAS Plan developments, in support of Development and Logistics.

Within FAA, two additional groups have explicit safety responsibilities.

- Aviation Safety. Reporting directly to the FAA administrator, Aviation Safety coordinates accident investigations, safety analyses, and special studies. Aviation Safety monitors safety activities of FAA programs, but does not function effectively as support to the operations of these programs.
- Mike Monroney Aeronautical Center. Located in Oklahoma City, the center houses the FAA Academy, the Civil Aeromedical Institute (CAMI), the Aviation Standards National Field Office, and the Airway Facilities National Field Support Center. The academy is the principal training facility for air traffic controllers. The Aviation Standards Training Branch at the academy trains flight standards and airworthiness inspectors, flight inspectors, and other personnel who work in Aviation Safety. CAMI researchers focus on improving selection and training for air traffic controllers, medically related aspects of aviation, including controllers' performance in the field, and physiological studies of pilot performance.

Other Federal Safety Roles

Other DOT offices oversee economic regulatory activities previously performed by CAB.

- The Office of the Secretary of Transportation (OST) issues certificates of public convenience and necessity required for all new carriers. OST also convenes government/industry meetings when necessary to handle scheduling peaks and delays.
- The Office of Aviation Operations and Aviation Enforcement and Proceedings in the General Counsel's Office performs fitness tests that examine a new carrier's management capability, financial posture, and regulatory compliance record.
- The Office of Aviation Information Management in the Research and Special Programs Administration collects economic information from major, national, regional, and commuter airlines as required especially under 14 CFR 241 and 14 CFR 298.

Although not a regulatory agency, the National Transportation Safety Board (NTSB) is an important institutional part of the safety structure. Created in 1966 as an arm of DOT, it became an independent executive branch agency in 1975. In addition to investigating commercial transport accidents, NTSB conducts special safety studies and issues recommendations that often call for rule revisions or for new federal regulations and procedures to correct safety problems. FAA conducts its own review of accidents and is not bound to accept NTSB suggestions for regulatory changes.

FAA Funding

Federal government funding for aviation-related programs comes from two sources: the Airport and Airway Trust Fund and from general tax revenues. The trust fund is financed by excise taxes on the aviation industry and its users, including an 8 percent ticket tax on commercial air passenger transportation within the United States. In addition, the unused portion of the trust fund accumulates interest credit payments from the Treasury. Currently, the largest contributor to the trust fund is the ticket tax, which accounted for 69 percent of the trust fund in 1985, followed by interest payments. Aviation excise taxes are deposited in the general fund of the Treasury. Although trust funds accounted for about 70 percent of FAA's total budget in fiscal year 1985, FAA consistently spends less out of the trust fund than is taken in from excise taxes and interest payments on the balance in the trust fund. Consequently money accumulates in the Treasury, where, according to current federal accounting standards, it can be counted against the federal deficit.[30] Critics of this policy believe the full resources of the fund should be available to FAA for operation and research and development rather than used as a tool to reduce the federal deficit figures.

Organizational Issues

System Safety Management. Notable in this brief description of FAA safety offices is the absence of a strong, internal system safety management advocate. A comprehensive approach to system safety could be described as:

> The application of engineering and management principles, criteria, and techniques to optimize safety within the constraints of operational effectiveness, time and cost throughout all phases of the system life cycle.[31]

Basic system safety management principles are applicable to commercial aviation and to the National Airspace System. A comprehensive system safety management program for FAA would apply to all aspects of planning, data collection and analysis, engineering, and operations. For example, the economic health and management stability of an airline strongly influence its ability and willingness to bear the cost of such safety activities as recurrent cockpit resource management and weather training for pilots, internal safety audits, and standardizing equipment and procedures. Yet while different offices within FAA have recognized the importance of all these factors, the agency has not systematized procedures to incorporate them in all areas of its oversight activities. Human error, the leading cause of commercial aviation accidents, also receives little FAA attention. These shortcomings speak to a need for coherent integrated safety management at FAA, beyond the development and enforcement of individual regulations and specific programs targeted at isolated problems.

In the absence of FAA system safety capability, this function is partially performed by groups such as Congress and airline labor unions, especially on issues where powerful interest groups differ vehemently (such as altitude encoding transponders). However, effective safety management is highly technical and requires continual close, objective attention to systemwide needs. These are beyond the capability of such groups.

System safety principles are also applicable to the NAS Plan, throughout all phases of its evolution and development of its elements, such as ATC technologies. NAS Plan programs often encompass some elements of system safety analysis. For example, the Traffic Alert/Collision Avoidance System (TCAS) program includes modeling and analysis of the effects of TCAS-induced maneuvers on air traffic, and other efforts to try to identify hazards in the use of TCAS before it is fully implemented. Procedural changes in the terminal area are evaluated through worst-case scenarios, operational judgment of experienced controllers, and other means, in an attempt to prevent accidents. These efforts are commendable, but maintaining or improving air safety as traffic levels increase will require a more systematic and broader approach to safety management. FAA's Office of Aviation Safety is developing system safety standards for FAA procurements based on military system safety standards. This is a good first step, but commitment will be needed to incorporate the principles fully into FAA's rigid technology development process, and, beyond that, into the entire life cycles of NAS.

The ATC system and supporting technologies warrant immediate special attention from a system safety perspective. The ATC system is currently under severe pressure to extend its operations to the limits of safe practice to meet the demand for service at busy hub airports. Furthermore, while the need to modernize ATC facilities is widely recognized, FAA's current plans include advanced automation features that are difficult to justify on the basis of efficiency and raise important human-factor questions. Rigorous system safety management, both for the near term and the longer term, would help maintain the excellent accident record of the ATC system, as FAA rises to the challenge of managing higher traffic levels. Resources are required for near-term ATC needs, such as increasing personnel levels and upgrading the computers in Terminal Radar Approach Control facilities. These are needed to accommodate increases in traffic and transponder users. Attention to formulating a better system safety groundwork for the more advanced parts of the system is also important.

Internal Communications. An additional and related problem is internal FAA communication paths. Vertical lines of communication exist between the administrator and the programs under the purview of the associate administrators and with the nine regional offices. However, the chart also illustrates that twenty-two separate groups report to the administrator and that no formal lines of communication are apparent among the operating programs and within program divisions. Moreover, even when communication lines exist, they are often ineffective because of timing and rigidity of responsibilities. For example, under the associate administrator for development and logistics, individual program managers in two offices are responsible for meeting milestones in the development and implementation of NAS subsystems. A third office is responsible for defining requirements and ensuring that individual subsystems combine effectively to form an overall system. Because many of the programs in the NAS Plan are already well underway by the time requirements are defined and validated, program managers have difficulty refocusing away from milestones and responding efficiently to inputs from other groups.

Within the last fifteen years, FAA has had seven administrators, serving an average of two years. Although this length of term is not unusual for administration appointees, this high rate of turnover highlights a central concern about FAA's capability to perform its safety mission—the requirement for long-range planning and policy commitment. Since many of FAA's responsibilities involve long-range programs, such as the modernization of equipment and facilities, the absence of consistent leadership is severely felt.

FAA REGULATORY PROGRAM (see box on page 45)

Although largely unnoticed by the traveling public, federal safety regulations, administered by FAA, establish the basic safety structure for U.S. aviation. Regulatory and oversight functions are primarily housed under the associate administrator for aviation standards, and activities of two of its offices are critical during times of major industry change.

The Office of Airworthiness

The Office of Airworthiness has two prime functions: to establish minimum standards for the design and manufacture of all U.S. aircraft and to certify that all aircraft meet these standards prior to introduction into service. Airworthiness standards prescribe explicit flight, structural, design and construction, powerplant, and equipment requirements.

The office issues "type"[32] certificates to prototype aircraft built

---

Overview of the FAA Regulatory Program

Federal Aviation Regulations are divided into two major parts:

- Part 121 applies to scheduled operations of large aircraft (more than 30 seats or a payload capacity greater than 7,500 lbs.)
- Part 135 regulates small aircraft (30 seats or fewer or a maximum payload capacity not over 7,500 lbs.)

The two principal regulatory programs are administered through nine FAA regional offices (see figure 3.1) and are directed by:

- Office of Airworthiness, which certificates aircraft and equipment.
- Office of Flight Standards, which governs all aspects of carrier operations.

The regulatory process has three components:

- Rule making to establish minimum standards, Federal Aviation Regulations.
- Inspections to certificate new carriers and to monitor compliance.
- Enforcement if noncompliance is found.

---

in conformance to airworthiness standards after successful testing. Manufacturers try to ensure that individual aircraft conform to the type to obtain FAA airworthiness certification. If major changes are made in an aircraft design, a new type certificate is required. However, if less extensive changes are made, FAA amends a type certificate and issues a supplemental one. As pilots must have additional and expensive training to operate a new type of aircraft, manufacturers and airlines prefer continuous supplemental certificates and pilot type ratings.

Four FAA regional offices have certification authority for aircraft and certain systems:

- Central Region (Kansas City) certificates general aviation aircraft.
- New England Region (Boston) certificates engine and propulsion systems.
- Northwest Mountain Region (Seattle) certificates large commercial aircraft.
- Southwest Region (Fort Worth) certificates helicopters.

This decentralized management lends itself to internal FAA disagreements over regulatory actions and sometimes outright contradictions.

Office of Flight Standards

Commercial aircraft are spot checked by flight standards inspectors to ensure they comply with Federal Aviation Regulations. This office certifies that new air carriers meet federal standards and approves flight procedures, determines some equipment regulations, and is responsible for seeing that inspectors conduct routine safety inspections.

FAA inspections are divided into three functional categories:

- Operations, including minimum equipment lists, pilot certification and performance, flight crew training, and in-flight record-keeping.
- Maintenance, including maintenance personnel training policies and procedures for overhaul, inspection, and equipment checks.
- Avionics, specializing in aviation-related electronic components.

Usually, each airline is assigned a principal inspector for each of the three categories of inspections. The principal functional area inspectors are assisted by inspectors from one of the ninety FAA district offices within whose boundaries the airline operates. In addition to certificating new airlines and performing routine inspections, FAA principal inspectors are responsible for investigation and enforcement duties.[33]

Regulatory Program Issues

In the years just prior to deregulation, standards and procedures followed by major U.S. airlines often exceeded minimum federal requirements. However, starting in 1978, economic forces exerted great pressure on redundancies in industry safety programs, eliminating some and intensifying the importance of strong federal enforcement programs. At the same time, FAA's capability to monitor the industry was swamped by problems, which were in part products of executive branch policies and governmental budget constraints, and which were independent of deregulation, although deregulation magnified their impact.

Investigations conducted since 1983 by FAA itself, the General Accounting Office (GAO), and NTSB cited weaknesses in the FAA inspection programs. OTA research confirms that severe difficulties persist, although work is underway to standardize procedures and provide for greater flexibility in personnel assignments.

Criticism of the FAA inspection program generally focuses on three categories: manpower and training, information systems, and management control. Manpower problems became acute during the early years of deregulation when federal budget constraints required cuts in the inspector work force. At the end of fiscal year 1978, FAA had 1,580 flight standards field office inspector positions authorized, and actual employment was 1,466. By fiscal year 1981, the authorization had risen to 1,748, and 1,615 inspectors were "on board" on September 30, 1981. Three years of deep budget cuts reduced the authorization by 18 percent to 1,440 inspectors by the end of fiscal year 1984. (Actual employment on September 30, 1983, was 1,331 inspectors.) At the end of fiscal year 1978, there were 556 "air carrier" inspectors employed (605 authorized) which increased to 623 (674 authorized) by the end of fiscal year 1981, and fell to 507 (569 authorized) by the end of fiscal year 1983. The planned end of fiscal year 1984 authorization was 508, later increased to 674 (the 1981 high). Thus, while the number of airlines was rapidly rising in the years following deregulation (the number of commercial operators roughly doubled between 1979 and 1983), the number of air carrier field inspectors in FAA was rapidly declining. Inspectors were shifted from routine operations and maintenance inspections to airline certifications. FAA's end-of-year goal for fiscal year 1988 is 2,088 field office inspectors, and FAA plans to add about 285 inspectors in each of fiscal years 1989, 1990, and 1991.[34]

Moreover, even if numbers of newly hired inspectors reach adequate levels, FAA inspector training programs cannot keep up with new industry procedures and equipment, such as contract maintenance work and new cockpit technologies. Training is most problematic in areas of recent technological development, such as advanced composite materials used by aircraft manufacturers, new navigational systems, and other computerized systems.[35] As aircraft and technologies become more complex and sophisticated, training for inspectors will become even more critical.

Furthermore, FAA managers have long lacked current and reliable information on allocation of inspectors and inspection records, leading to inconsistencies among FAA district offices, and inadequate followup to inspection activities. Shortages of computerized equipment and lack of high-quality core training at the Oklahoma City academy exacerbate information difficulties.

Traditionally, FAA has delegated broad authority to regional and district offices concerning the frequency and scope of inspections. FAA regional offices stoutly reaffirm the importance of meeting regional needs at the regional level, leaving general policy guidance to Washington. However, FAA headquarters has never effectively centralized management control to permit evaluating regional and district inspection activities, to ensure uniformity in policies and procedures, and to analyze inspection findings on a national scope. Wide variations in the number and kind of inspections performed from region to region identified by GAO in 1985, still persisted according to OTA's research.[36]

The competence and professionalism in the manufacturing and operating industries ensure airworthiness of commercial aircraft to current standards, and past history shows that industry safety standards are almost always high. FAA needs adequate technical expertise and records to be able to target the rare cases where standards are not sufficiently high, as well as knowledge about industry management attitudes and financial stability.

To improve management of its inspection responsibilities over the long term, FAA initiated Project SAFE, a program to establish staff standards, increase staff levels, improve inspector manuals and training courses, and establish performance standards for each FAA regional office. Task forces made up of headquarters and regional staff are revising and standardizing inspection manuals and training policies. Needed improvements to training courses in Oklahoma City and standardizing of regional on-the-job training are planned under Project SAFE, but are moving at a snail's pace. Moreover, emphasis on monitoring individual airline characteristics, such as compliance records, fleet composition, management changes, and financial stability, would permit FAA to allocate its inspector resources more effectively.

Adequacy of FAA Minimum Standards

Although most airlines maintain standards above the minimum required by FAA, some safety officials are concerned that the minimum may not be adequate in some instances. Because of such concerns, the Department of Defense has instituted a safety program that frequently uses a higher standard in selecting contract airlines than the minimum standards required by FAA.

In response to the 1985 crash of a military chartered DC-8 in Gander, Newfoundland, the Military Traffic Management Command (MTMC) and the Air Force Military Airlift Command (MAC) overhauled their inspection program and established an Army/Air Force Central Safety Office to coordinate standard setting and inspection activities. Enforcement actions against the airlines are the responsibility of a military review board.[37] The MTMC/MAC office conducts inspections, in addition to FAA's, of all airlines used for military charters. During the two years since the Gander crash, the safety office has disqualified thirteen U.S. airlines and taken lesser disciplinary actions against nine others. Poor maintenance practices and failure to comply with airworthiness directives are the most frequent problems. Half of the cited airlines were large carriers operating under Part 121.

By the summer of 1988, the MTMC/MAC safety program will be supported by a new database. The Air Carrier Analysis System (ACAS) will compile and analyze data on airline accidents, incidents, maintenance and operating problems, and financial characteristics. The system will alert inspectors to those circumstances at an airline that warrant personal inspections and provides a useful model for FAA, which is cooperating with MTMC/MAC.[38] However, ACAS relies upon FAA databases which are incomplete and are not designed to support analyses.

The FAA Rule-Making Process

Prior to deregulation, FAA had considerable regulatory autonomy, overseeing an industry in which profits were protected through the extensive rate and entry rules of CAB. Over the past decade, vigorous industry economic competition has made rule making a distinctly adversarial process. Carriers, labor groups, aircraft manufacturers, and general aviation supporters carefully scrutinize every proposed safety regulation and question its efficacy and impact on costs. Often such activities, in concert with administrative policies and bureaucratic labyrinths, have effectively blocked safety regulations for years.

Presidents Ford, Carter, and Reagan each initiated progressively stronger and more centralized programs of regulatory review in response to concerns about the excessive burdens and inadequate management of federal regulations. These policies, implemented explicitly through executive orders in 1981 and 1985, direct agencies to:

- base their regulatory rule-making decisions on benefit-cost analyses;
- submit new regulations for review by the Office of Management and Budget (OMB);[39]
- refrain from starting work on any significant new regulation until consulting with OMB; and

- publish in the annual *Regulatory Program* a status report on each significant regulatory initiative.

While all executive branch agencies have had to revamp their regulatory procedures as a result of these executive orders, FAA has faced a special challenge because proposed remedies to safety risks often entail expensive technological developments requiring long lead times.

Moreover, DOT has gone substantially beyond executive order mandates for economic review of proposed rules for all its modal agencies. Cost-benefit analyses are required only for identified "major" regulations, but in contrast to some other executive branch agencies, DOT expanded this requirement to include "significant" rules, a category that covers nearly all regulations.[40]

For FAA, the review process now consists of the following major steps:

- FAA advises the Office of the Secretary of intent to start work on a significant regulation. DOT departments register concerns about the proceeding or about analysis needed.
- A team of FAA staff members develops a new rule proposal.
- A member of FAA's Office of Aviation Policy, who also serves on the rule-drafting team, prepares the cost-benefit analysis.
- After FAA approval, the regulatory package, complete with economic analysis, moves to DOT's General Counsel Office for a required departmental review, including assistant secretaries for Policy and International Affairs, Government Affairs, and Budget and Programs.
- Prior to public release, the general counsel mediates OMB's review of the regulation and economic analysis.

In a major review of its regulatory program in 1984, FAA identified over one hundred regulations needing revision. Twenty-six regulations were assigned high priority status and are currently in various stages of the process; another eighty-five form a large backlog. Long backlogs can lead to "immediate action" regulations and inspector handbook changes that alter regulations without adequate due process. While FAA plans a rewriting of Part 121 and 135 regulations, this major undertaking will require years of intensive effort.[41]

NOTES

1. Public Law 85-726.

2. Nick A. Komons, *Bonfires to Beacons* (Washington, DC: U.S. Department of Transportation, Federal Aviation Administration, 1978), p. 25.

3. Initially, air mail contractors were paid a percentage of postage revenues. In 1926, however, an amendment to the Air Mail Act of 1925 required payment by weight carried.

4. Key issues debated by Congress included whether to separate military and civil aviation activities, what responsibilities should be left to state and local governments, and how to provide federal support for airports. Komons, *Bonfires to Beacons*, pp. 35-65.

5. *Congressional Record*, 69th Cong., 1st sess., May 20, 1926, 9811.

6. R. E. G. Davis, *Airlines of the United States Since 1914* (Washington, DC: Smithsonian Institution Press, 1972), p. 201.

7. Komons, *Bonfires to Beacons*, p. 124.

8. Pilots were restricted to flying 100 hours per month, 1,000 hours during any 12-month period, 30 hours for any 7-day period, and 8 hours for any 24-hour period; a 24-hour rest period was also required for every 7-day period. These requirements, established in 1934, and virtually the same today, upgraded earlier restrictions which limited pilots to 110 hours of flight time per month. In addition, a waiver of the 8-hour limitation for a 24-hour period could also be granted by the Department of Commerce. The 8-hour waiver rule was ultimately eliminated following a fatal accident involving a pilot who had exceeded 8 hours of flight, and pressure from the Air Line Pilots Association. Ibid, pp. 290-292.

9. Ibid, pp. 116-118, and Davies, *Airlines of the United States*, p. 201.

10. The fatality rate rose from 4.78 per 100 million passenger-miles in 1935 to 10.1 per 100 million passenger-miles in 1936. Komons, *Bonfires to Beacons*, p. 295.

11. Civil Aeronautics Act of 1938, Public Law 75-706.

12. Increasing air traffic between Newark, Cleveland, and Chicago prompted a group of airlines to establish an air traffic control system in 1934. By 1936, however, the Department of Commerce assumed control of the system and issued new regulations for instrument flight. Komons, *Bonfires to Beacons*, p. 312.

13. John R. M. Wilson, *Turbulence Aloft: The Civil Aeronautics Administration Amid Wars and Rumors of Wars, 1938-1953* (Washington, DC: U.S. Department of Transportation, Federal Aviation Administration, 1979), p. 152.

14. Ibid, p. 261; and 43 *Federal Register* 46734 (Oct. 10, 1978).

15. Wilson, *Turbulence Aloft*, p. 161.

16. The Civil Aeronautics Board adopted 14 CFR 298, designating an exempt class of small air carriers known as "air taxis."

17. Initially, Lockheed produced the Constellation which carried sixty passengers and was seventy mph faster than the DC-4. To compete with Lockheed, Douglas developed the DC-6. Subsequently, upgraded versions of each aircraft—the DC-7 and the Super Constellation—were introduced. Davies, *Airlines of the United States*, p. 289.

18. The four carriers were Air News, Flying Tigers, Slick, and U.S. Airlines. See Nawal K. Taneja, *The Commercial Airline Industry* (Lexington, MA: Lexington Books, D. C. Heath and Co., 1976), p. 6.

19. 44 *Federal Register* 66324 (Nov. 19, 1979).

20. Public Law 85-726, Aug. 23, 1958, 72 Stat. 731.

21. However, the Federal Aviation Administration administrator was authorized to play an appropriate role in accident investigations. In practice, the Federal Aviation Administration routinely checked into accidents for rule violations, equipment failures, and pilot errors. Moreover, the Civil Aeronautics Board delegated the responsibility to investigate nonfatal accidents involving fixed-wing aircraft weighing less than 12,500 pounds to the Federal Aviation Administration. Stuart I. Rochester, *Takeoff at Mid-Century: Federal Aviation Policy in the Eisenhower Years, 1953-1961* (Washington, DC: U.S. Department of Transportation, Federal Aviation Administration, 1976), p. 234.

22. 49 U.S.C. 142(a).

23. See 49 U.S.C. 1430, 1422-1424, 1426, 1427, and 1432. Procedures for amending, suspending, or revoking certifications are contained in 49 U.S.C. 1429.

24. Federal Aviation Administration staff grew from 30,000 in 1959 to 40,000 in 1961. Rochester, *Takeoff at Mid-Century*, p. 295.

25. U.S. Department of Transportation Act, Public Law 89-670, 49 U.S.C. 1651.

26. For additional information on implementation problems associated with the trust fund, see J. Glen Moore and Patricia Humphlett, Congressional Research Service, "Aviation Safety: Policy and Oversight," Report No. 86-69SPR, May 1986.

27. U.S. Department of Transportation, Federal Aviation Administration, *FAA Handbook of Statistical Information, Calendar Year 1986* (Washington, DC: 1987), p. 118.

28. Public Law 95-504, 92 Stat. 1703. It was thought that fares would drop based on the record of intrastate airlines where fares were 50-70 percent of the Civil Aeronautics Board-regulated fares over the same distance. In addition, the Civil Aeronautics Board had already reduced restrictions on fare competition in 1976 and 1977 and allowed more airlines to operate in many city-pair markets. Robert M. Hardaway, "Transportation Deregulation (1976-1984): Turning the Tide," *Transportation Law Journal* 14, no. 1 (1985): 136.

29. Civil Aeronautics Board Sunset Act of 1984, Public Law 98-443, Oct. 4, 1984, 98 Stat. 1703.

30. U.S. Congress, General Accounting Office, *Aviation Funding: Options Available for Reducing the Aviation Trust Fund Balance,* GAO/RCED-86-124BR (Washington, DC: U.S. Government Printing Office, May 1986).

31. U.S. Department of Defense, *Military Standard: System Safety Program Requirements,* MIL-STD-882B (Washington, DC: Mar. 30, 1987).

32. 14 CFR 1.1 (Jan. 1, 1987).

33. U.S. Congress, General Accounting Office, *Report on Aviation Safety: Needed Improvements in FAA's Airline Inspection Program are Underway,* GAO-RCED 87-62 (Washington, DC: May 1987), p. 12.

34. Anthony J. Broderick, associate administrator, Aviation Standards, Federal Aviation Administration, personal communication, Mar. 31, 1988.

35. General Accounting Office, *Report on Aviation Safety,* p. 50.

36. U.S. Congress, General Accounting Office, *Compilation and Analysis of the Federal Aviation Administration's Inspection of a Sample of Commercial Air Carriers,* RCED-85-157 (Washington, DC: Aug. 2, 1985); and OTA primary research.

37. James Ott, "Military Avoids U.S. Carriers That Fail Safety Standard," *Aviation Week and Space Technology,* Feb. 8, 1988, pp. 99-101.

38. Ibid; and Broderick, personal communication.

39. Thomas Hopkins, "Aviation Safety Rulemaking," OTA contractor report, August 1987, p. 5.

40. Ibid, p. 7.

41. General Accounting Office, *Report on Aviation Safety,* p. 26.

2

High Risks, Sinking Fortunes*

The business of making and selling commercial airliners is not for the diffident or faint of heart. It is remarkably difficult and, by anyone's standard, intensely competitive. There are a few industries that consume as much or more capital; certain others rely as heavily on quantities of highly skilled personnel; probably no other is involved with as many advanced technologies.

But what really sets the commercial airplane business apart is the enormity of the risks as well as the costs that must be accepted; they create an array of obstacles to profitability, hence viability, which discourages all but the bold and committed. A strong current of zeal runs through this industry, which is operated in a free-wheeling, bare-knuckled competitive style. And while the fee for entering the competition is injuriously high (a new airplane program will devour $2 billion long before deliveries begin), the pro-

---

* Reprinted, with editorial changes, from *The Sporty Game*, by John Newhouse, Copyright © 1982 John Newhouse, by permission of Alfred A. Knopf, Inc.

cess itself is exciting and the rewards, if attainable, are high and include power and influence on a world scale. Hence, many have tried, few successfully.

The turbulent do-or-die, all-or-nothing environment in which airliners are made and sold is reminiscent of the nineteenth century and some of the entrepreneurial spirits of that era. In deciding to build a new airliner, a manufacturer is literally betting the company, because the size of the investment may exceed the company's entire net worth. The remarkable scale on which they operate induces a curiously understated, rather casual style among senior executives in this industry. Betting the company, for example, is being "sporty." Failure comes swiftly, seemingly overnight. A few key airlines make or break an airplane program: if there is a choice of two new airplanes, these airlines may select one, thereby extinguishing the other; or they may diverge and split a usually limited market by buying both airplanes, in which case the competing suppliers are driven to accept prices that don't come close to covering their costs.

By contrast, the auto industry's mass market has sustained a wide variety of suppliers and innumerable models over periods of many years,excusing cars that haven't suited the motorists' larger interests and the gaffes of their makers. The unfortunate Edsel stands out as a dramatic exception to the forbearance of America's car market; that debacle is judged to have cost the Ford Motor Company about $350 million. And yet the company hardly flinched and continued profitably marketing the rest of its line. This is the sort of misfortune that occurs naturally and regularly in the airliner business, where the effects are unavoidably grave or catastrophic.

Sometimes a manufacturer that bets on a new airplane and loses will exit from the commercial business along with the failed product. The Glenn Martin Company, General Dynamics, and most recently Lockheed, among others here and in Europe, are in that category. And for every type of airliner actually produced and sold to the airlines, scores of other "paper" airplanes die in the larval, or brochure, stage. Starting in 1952 with Britain's de Havilland Comet, an airplane prone to metal fatigue and hence disaster (two of them fell apart in the air), there have been twenty-two commercial jet-powered transports, of which only two, thus far, are believed to have made any money. These are the Boeing Company's first two entries: its long-range 707 and the medium-range 727, the industry's biggest seller. Boeing's largest airplane, the 747, may one day be judged to have crossed the break-even point, whereas the company's smallest, the 737, although it may eventually approach the 727 in sales, is not likely to cross the break-even line. The McDonnell Douglas Corporation's DC-8 probably would have been profitable but for some questionable decisions by management, especially the decision to kill it in favor of the DC-10; the firm's small two-engine DC-9, which preceded its chief rival, Boeing's 737, is unlikely to make any money, and the troubled DC-10 looks like a certain money loser. Lockheed's L-1011—an airplane that is more admired within the industry than most others, including its rival, the DC-10—had lost $2.5 billion by the time it was canceled.

In varying degrees, these estimates are impressionistic; they reflect a consensus judgment within and around the industry. The truth is, nobody knows where the break-even point of a commercial airplane program lies, nor can anyone price an airplane from the cost of making it. Exact figures are obscured by a monumental set of variables and imponderables. Malcolm Stamper, Boeing's former president, says, "Locating the break-even point is like finding a will-o'-the-wisp." He adds that

> setting the price of an airplane is not as difficult as measuring the break-even point. The airplane has to be competitive. Thus, it prices itself—not as a function of actual costs but of the competition and in terms of what it does for the customer [the airline] in passenger-seat-mile costs and ton-mile costs.

And Richard W. Welch, the president of Boeing's Commercial Airplane Company, answers the question about where the break-even point should lie with another question: "What are we talking about? It's very hard to define." On a related matter, Welch is more certain. "The risks," he says flatly, "are greater in this business than in any other compared to the returns."

Until recently, those who have succeeded were American firms, particularly the Douglas Aircraft Company—now a division of the McDonnell Douglas Corporation—and the Boeing Company, the industry's current colossus. However, matters are not as they were: a number of developments occurring over the past fifteen years or so, including a remarkable series of major miscalculations and mistakes committed by the American firms in this business, have quite unexpectedly created a situation in which Americans are pitted less against each other, as in the past, than against a Western European consortium called Airbus Industrie, which produces a twin-engine, double-aisle airplane known as the Airbus that is highly regarded wherever it is flown. This transatlantic competition seems to challenge America's traditional dominance of a major industry built around the technologies on which much of economic strength and growth is judged to depend. Probably no other industrial products embody as much refinement in the combined techniques of metallurgy, electronics, and computers as do jet aircraft.

Apart from data processing, the only so-called high-technology industry in which America is holding its own in the world market is aircraft, most notably commercial aircraft. The celebrated technology gap of the 1960s between the United States and the rest of the industrialized world has vanished, and there is reason to doubt that it ever really existed. Patent awards related to industrial technology are increasing sharply in Europe and Japan, far more so than in the United States. American corporations have become less innovative and less productive than many of their foreign competitors, who outperform them in a variety of industries, including automobiles, steel, machine tools, plastics, and consumer electronics. European and Japanese export industries are not only fully competitive technologically with America's, to say the least, but can rely on financial support from their commercial banks, which are highly responsive to the export policies of their governments. American banks make their own policies, and many American exporters are having serious difficulties in the capital markets. Access to capital, however, is not among the problems of American aircraft suppliers, in part because they still hold the major share of the world market, a state of affairs that also helps to explain the country's special dependence on this industry. The sales in foreign markets of jet transports and engines are relied upon to relieve pressure on America's economy arising from deficits in its international account; indeed, sales of these airplanes abroad have for many years earned more foreign exchange than any other U.S. export. Over the past five years, 67 percent of airliner sales have been abroad.

There is a strong possibility, however, that the present performance of Airbus Industrie, if sustained, will help to drive one of the two remaining American competitors—McDonnell Douglas—from the commercial market. The American companies have reacted with the indignation that comes naturally to established institutions when confronted by *arrivistes* and usurpers. Industry's outcry has echoed loudly in numerous congressional offices and in the corridors of federal agencies which deal with the commercial airplane industry. The concern, essentially, is that Airbus Industrie has direct government backing, an advantage which, if fully exploited, might in the end be decisive. Actually, the Airbus consortium consists of several government-controlled companies, of which two—one in West Germany and another in Britain—have some private ownership. The funding comes chiefly from governments, particularly the French, West German and British, and key decisions, such as whether to build a new airplane or to sell airplanes at a sizable loss in order to win a competition, are taken by governments. In the presumed determination of European governments to sustain Airbus Industrie and promote its fortunes in the world market, worried Americans see a unique national asset—the commercial airplane industry—being whittled down, just as other of America's prize industrial assets have been depleted by foreign competition.

In its extreme form, the concern is that Airbus Industrie threatens even Boeing in its role of major supplier of airliners to the world market, a market that is growing most rapidly neither in Europe nor in the United States but rather in Asia, the Middle East, and Latin America—regions where no one supplier has a reliable competitive advantage. The market for new airplanes over the next decade should exceed $150 billion. In contemplating this apparent bonanza, the American anxiety extends beyond Europe's potential. People worry about which side, if either, the Japanese will choose as a partner when they decide, as they surely will, to broaden their modest but growing airplane-building capability; both Americans and Europeans who have observed this capability use words like "awesome" and "fantastic" to describe it.

The transatlantic competition in commercial aircraft has released tendencies, some of which are undesirable, some probably desirable, and some in conflict with each other. Each side's industry, for example, is working strenuously to protect its domestic market from the other's competitive wares, a tendency that may cause a destructive waste of valuable and limited resources and lead to a kind of formalized self-sufficiency, or autarchy, that could disturb relations between America and its principal allies. And each side is using whatever tactics it can devise to obtain the largest possible share of the Third World market. But for wholly practical reasons, each side is also blurring these tendencies by acquiring components of the other's industry as partners and subcontractors. The idea is that an airplane that is multinational in its parts and fabrication offers the seller political advantages in the world market; all of the major suppliers and their governments see a need to dilute the national character of their products. They also know that one company can no longer take on the full burden of a commercial airplane, or an engine for that matter. Any supplier who launches a new air-plane or engine program must involve innumerable other compa-nies, large and small, which are spread around the major industri-alized nations of the West and now Japan as well. A fierce competition for risk-sharing partners and subcontractors is under way. Governments are drawn in. Presidents and prime ministers become involved. Briefly, the political stakes in the airliner busi-ness are rising along with the costs and the risks.

Because the risks are so great, there has been a tendency, though one of necessarily declining force, to take them for granted. Joseph Sutter, one of the world's preeminent designers of large air-planes, says: "In this business, you have to put the company on the line every three or four years." Sutter works for the Boeing Com-mercial Airplane Company, which now accounts for about three-fourths of the annual earnings of its parent, the Boeing Company. His title is executive vice-president, and for years he has supervised both the design and the development of new airplanes and their fab-rication. This amounts to having direct responsibility for turning out not just one or two at a time but a so-called family of commercial airplanes. Boeing's chief competitive advantage lies in its unique capability to offer a large and growing assortment of airplanes—wide-bodies and narrow-bodies, some with two engines, others with three and four. More and more, the airlines of the world require variety in their fleets—airplanes of varying size and range for use on routes of varying length and passenger density. Moreover, in the airplane business, as elsewhere, nothing succeeds like success. Boe-ing makes more than half of the jet-powered airliners sold on the world market, which includes virtually all countries outside the Soviet bloc. "We are good," Sutter says, "partly because we build so many airplanes. We learn from our mistakes, and each of our air-planes absorbs everything we have learned from earlier models and from our other airplanes."

He could also have cited Boeing's astonishing productivity. There are several explanations for Boeing's ability to assemble air-liners faster than any competitor in the world, and while there is probably some truth in each of them, the most entertaining is John G. Borger's. Borger, a legendary figure, was for many years Pan American's vice-president for engineering. He has worked for Pan Am since graduating from the Massachusetts Institute of Technology in 1934, and he was heavily involved in the creation of more than a dozen airliners. Borger, a native New Yorker, recently retired from his troubled company. He has a very good and reliable memory. He recalls that when Pan Am became the first, or "launch," customer for the 707, the first jet airliner, Juan Trippe made a bet with William Allen, his opposite number at Boeing. (There is a lot of wagering between airlines and their suppliers.) This bet was that Boeing could not deliver Pan Am's 707s ahead of schedule, but that for each month any of them actually was delivered early Trippe would pay a bonus of $25,000. Borger says that Allen won the bet, but doing so, according to some Boeing people, probably cost his company $300,000 per airplane. However, many people, Borger adds, are sure that Boeing's productivity, which constantly improves, is at least partly traceable to the pressure set off by the Trippe-Allen bet. When asked about the bet a year before his death in April 1981, Trippe had forgotten it. But Allen, now in his eighties and honorary chairman of Boeing, having retired in 1972 after forty-seven years with the company, remembers it. He still maintains an office at the corporate headquarters south of Seattle and is a revered figure at Boeing. Malcolm Stamper says: "We're all picking apples from Bill Allen's tree." Although he doesn't recall the exact details of the bet with Trippe, Allen says that it was the "biggest I ever made." And, he adds, "it did help productivity." He doesn't agree that the wager cost Boeing money. To the contrary, he says, "we needed every nickel we could get."

Today, Boeing is better positioned to cope with the rigors of the commercial business than its competitors. However, Boeing, which once relied almost entirely on the bombers and other weapons systems it sold to the Pentagon and had withdrawn badly scarred from the commercial field, is as vulnerable as any competitor to its hazards. It is not enough to build and market competent jet airliners, even a family of them. At some point, some of the airplanes must return a profit. And at a time when immensely powerful engines, along with a stream of refinements, have made it possible to offer air travelers more reliable airplanes, the business of making and selling them has become even more onerous. The heaviest of the American suppliers' burdens is the weakness of their domestic market—the airline industry of the United States. Its problems not only compare with some of those which afflict suppliers but resemble them; that is hardly surprising, since the two industries function in tandem and have to find answers jointly to many of the same hard questions. When either of them sneezes, the other catches cold.

During the past thirty years or so, the airlines of the United States became one of the major growth industries, but an especially vulnerable one. Probably no other industry has endured as many ups and downs—as much "cyclical shock," as it is called. The volatility of the airlines' finances is reflected in their earnings and stock prices. In the early 1960s, the average price of airline stocks was a little over $5; by 1966, it had soared to $47, but in 1970 had fallen by 75 percent to $13. The major, or trunk, airlines earned over $1 billion in 1978 and $400 million in 1979, and *lost* $225 million in 1980, the first year on record during which passenger traffic actually declined; in 1981, the loss rose to $300 million. In bankers' language, the airlines—most of them—are highly "leveraged," meaning that the debt-to-equity ratio in the airline industry is high, higher than in most others. For these and other reasons, airlines are not popular with banks and other lending institutions, which is troublesome, because, like their suppliers, airlines are large borrowers. Managing an airline is one of the more inexact sciences. During much of the history of the industry, its pioneers, many of them former pilots, lingered on as dominant figures. A few of these, such as Pan American's Juan Trippe, were always equal to the prodigious roles they created for themselves; but at least as many others did not adjust to the rapid growth of their trade and the increasing variety of its problems. The aura of silk scarves and leather puttees disappeared from the corporate front offices years ago, but the management of many airlines is viewed skeptically by the watchful lenders and the investment banks.

The airlines, by and large, are micawberish and stubbornly unrealistic. They didn't understand what lay behind the soaring growth in air travel in the 1960s, their golden era. Much of it came at the expense of railroads and steamships; thus, even when the economy was in a slump, air travel didn't fall off as long as the shift from trains and ships to airplanes was still under way. By the end of the 1960s, however, the transition was largely complete, which meant that once again the fortunes of the airlines could be expected, at best, to rise and fall with the economy. The recession of the early 1970s, by halting the growth in passenger traffic, reacquainted the airlines with this harsher aspect of their business.

A few of them—Delta Air Lines is a paramount and continuing example—are demonstrably well run and have usually made money in the down as well as the up periods. But most of the carriers are struggling while their troubles intensify. Buying the airplanes and the engines is normally the hardest problem, intellectually as well as financially, for an airline; it is the most fateful decision that management and the directors can make. The airline must live with the equipment for fifteen to twenty years or more. Harry E. Colwell, vice-president for aerospace of the Chase Manhattan Bank, relates easily to the problem as it appears to airline chief executives. "How can you know," he asks, "if you are running an airline, what your operations will be like in a few years' time? Which routes will you be flying? How productively? What stage lengths [distances between cities]? Against what kind of competition? You are buying ahead and, in fact, are so far out in the future that you can't really be sure you know what you are doing."

Buying an airliner is like buying a car only in the sense that the airline customer, like the motorist, is offered lots of options; the supplier will do virtually anything within reason to be accommodating. Actually, it more closely resembles buying a house, because most airlines keep their airplanes in service for fifteen years or so before acquiring something newer, and they replace them, not so much because they wear out—airliners don't wear out easily—but because requirements change. (Obviously, newer airplanes also offer improved efficiency and marketing advantages.) An airline must try to calculate its future needs just as a family does when buying a house; the airline, too, must avoid saddling itself with more equipment than it can manage financially. Picking the right airplanes and finding the right balance between needs and resources is all very difficult. Kenneth G. Wilkinson, who once ran British Airways and is among the most admired figures in European aviation, says that in selecting equipment, "however brilliant you are, you never get it exactly right, but you can't afford to get it too far wrong." The problem confronts the suppliers with equal force. Each of them, if it wishes to stay in the game, must choose the right moment to bet the company by launching the airplane, or airplanes, which, according to its calculations, will best meet both the short- and long-run needs of the airlines.

Cars have grown smaller and commercial airplanes larger, in both cases because of relentless pressure for improved economy. The airlines resemble, and even think of themselves as, conveyor belts which are constantly moving buckets (the airplanes) containing sand (the passengers and baggage). Increasing the size of the buckets, theoretically, will lower operating costs: as the number of people an airplane can carry rises, the cost to the airline of transporting each of them individually—the so-called passenger-seat-mile cost—goes down, provided, that is, enough seats are actually filled. The percentage of seats on a given flight occupied by passengers is known in the airline industry as the load factor and is, naturally, a basic measure of profit and loss.

The customary way to expand an airplane's revenue-earning capability is to build a stretched version; stretching the long oval hull of a modern transport aircraft to create space for more seats is a relatively simple matter aerodynamically. However, an airplane can be stretched only so far. Hence there are moments in the striving by airlines for greater "productivity" when an option to replace existing airplanes with new and larger—and sometimes faster—ones appears and instantly acquires important partisans. In the mid-1960s, the manufacturers of the airplanes and their airline customers became jointly fascinated by the possibilities of wide-bodied, double-aisle airplanes; jumbo-sized vehicles of this kind seemed not only feasible but capable of doubling the productivity of even stretched versions of existing transport aircraft, thereby making possible deep reductions in direct operating costs. Yet, building a wholly new airplane is far more costly than stretching one that already exists. And the jump from single- to double-aisle airplanes—from big to jumbo—would clearly be reflected in the far heavier costs to all parties. Even more sobering, wide-bodied commercial airplanes would require much larger and more powerful engines of a kind that were not yet available and would further escalate the price to the airlines of equipment so novel. (At that, the wide-bodied airplanes were considerably less novel than the supersonic transports that most of the major airlines in the United States and elsewhere were also looking ahead to and planning to buy.)

Airline chief executives and their bankers could justify the cost and risks if they assumed that a period of substantial, uninterrupted growth lay ahead. Anyone who questioned that dubious assumption clearly had no business contemplating airplanes larger than the stretched version of Boeing's 727, in the medium-size and range category, and the stretched version of Douglas's DC-8, in the larger and longer ranges. However, nearly everyone was unreservedly optimistic, or professed to be. The seers in industry and in government agencies were all forecasting a steady annual increase in traffic of about 15 percent. Growth on that scale, in turn, would offer airline managers the promise of load factors high enough to assure profitability and eventually vindicate a gamble on wide-bodied airplanes. Most of the world's major airlines did order them, even though passenger traffic in many countries, the United States in particular, has not since achieved anything resembling significant growth over a span of even a few years.

In effect, both the carriers and their major suppliers got it wrong by exaggerating. Too many different wide-bodied airplanes were launched for a market that was smaller than either the trunk airlines or the suppliers themselves realized. The carriers encouraged their suppliers to create the huge new airplanes, and then simply bought too many of them. Suppliers and carriers alike ignored the harsh lessons of their experience, and became hostages to a future which their forecasts painted in pure hyperbole. Two of the three major suppliers, the Boeing Company and the Lockheed Aircraft Corporation, were dragged, mainly by the new programs, to the edge of financial ruin and collapse. The third, the McDonnell Douglas Corporation, at first had fewer and less-severe problems in the new era, but its own wide-body, the DC-10, gradually became a problem that has proved lingering, painful, and injurious to the company's reputation and its earnings.

Exaggeration weakened the resistance of the airlines to the cyclical shocks that were not supposed to occur but lay ahead. The worst of these was the dramatic rise in fuel prices that began after the Yom Kippur War in October 1973; both the airlines and the suppliers were thrown far off stride by a development which the politics of oil had made inevitable and predictable, with or without a war in the Middle East. The effect has been a steeply rising percentage of the airlines' direct operating expenses represented by fuel costs. In 1973, the cost of their fuel to the airlines was eleven cents per gallon and amounted to about 20 percent of their direct operating costs. By 1981, the cost per gallon of kerojet, the kerosene-based fuel burned in jet aircraft engines, was above a dollar per gallon and amounted to nearly 40 percent of direct operating costs. There is little the airlines can do to lighten this heavy burden, other than to share it with passengers by raising fares, a practice now increasingly adopted. On many flights which combine high frequencies with high passenger density (load factors), such as the Eastern Air Lines shuttle service which links New York, Boston and Washington, fares have been regularly raised to offset increases in fuel prices.

However, nothing in the airline industry is simple. The device of linking fares more directly to rising costs has collided with a parallel practice of slashing fares in order to try to cope with the industry's two other major problems—excess capacity and excess competition—both of which are self-inflicted and rich in precedent. Talleyrand's uncourtly appraisal of the Bourbons—"Nothing learned, nothing forgotten"—captures the airlines: they seem to follow a self-renewing cycle which exposes them to their periodic shocks. In good times, larger and perhaps faster airplanes are purchased. The new equipment permits—indeed, dictates—expanded service, the predictable effect of which is increased competition, with more airlines contesting each other on more routes. (Or, in not-so-good times, there is often one carrier which will try to seize a competitive advantage by buying a new airplane; the others, however squeezed, usually queue up for the same equipment.) Before long, the larger airplanes and the expanded competition have created a capacity that exceeds the needs of the market—the people who travel by air. At that point, pressure develops to increase some fares but also to lower others. Larger revenues on some routes are supposed to offset declining load factors, as well as costlier fuel and the country's highest labor costs, while lower fares on other routes are intended to sharpen a carrier's competitive edge and generate more traffic. Whether those ends are achieved or not, the reductions do have the perverse effect of raising break-even load factors and thereby weakening, if not altogether nullifying, the benefits to an airline of the fare increases. So it is that passengers on high-frequency flights lasting less than two hours may be paying as much or more for their tickets as many travelers flying the less frequent transcontinental routes—New York to Los Angeles, say—for which the competition is far more intense.

Between the spring of 1979 and the spring of 1980, the number of airlines competing for the New York-California market rose from three to seven. A senior executive of Pan American World Airways, one of the new arrivals in this market, observed, "When everyone shifts to the same profitable route, they all have the pleasure of losing money together." He was thinking not just of the excessive competition but of the senseless fare wars that it sets off. The carriers know that these are invariably self-destructive; but they can't restrain themselves. Like spent boxers, they answer the bell if they can. Jack S. Parker, who recently retired as vice-chairman of the General Electric Company and worked closely for many years with the airlines while directing General Electric's aircraft engine business group, says: "I just don't know of an industry that is more dedicated to self-liquidation than the airline industry, the way they handle their fare problems. The minute they get an increase in their fares, then they've got some new schemes to try to increase traffic at lower fares."

Most of the trunk carriers have a stubborn preference for the long-distance express over the local. The express is a much simpler operation; typically, a wide-bodied airplane carrying several hundred passengers climbs swiftly to a cruising altitude, where it can be flown most economically, and remains at that altitude for several hours until it nears its destination. The shorter and higher-frequency flights repetitiously involve an airplane in the high-cost activities of climbing to and descending from cruising altitudes; also, increasing congestion in the air lanes and airports strikes at the profitability of the shorter, high-frequency flights by causing delays in arrivals and departures.

In commerce as in life, however, the harder path is often the more reliable way to success. The air transport business is largely and unavoidably devoted to short-haul, high-frequency flights—to the hub-and-spoke route structures as they are called. Close to three-fourths of all scheduled flights worldwide are of less than two hours' duration and less than a thousand miles in length. The average flight within the United States is six hundred and sixty-five miles. The workhorses of the world's airlines are the middle-sized and smaller jet transports, mainly the Boeing 727s and 737s and McDonnell Douglas's DC-9s. The larger of the widebodied airplanes—Boeing's 747, with four engines, McDonnell Douglas's DC-10 and Lockheed's L-1011 TriStar with three—cannot be operated profitably on the shorter routes because of their size. As for passenger preference, most of the people who regularly travel the shorter routes care much less about the interior dimensions of the airplane than they do about the convenience of having an assortment of departure times from which to choose.

JOHN FIELDER

3

Floors, Doors, Latches, and Locks

IMPORTANCE OF THE CARGO DOOR LATCH AND LOCK SYSTEM

It is important to have a reliable mechanism to close the aircraft doors (latch) and another to make sure that they stay latched (lock). Besides the obvious danger of loss of passengers or cargo if the door opens in flight, there is the possibility that lost cargo or the detached door may damage the tail of the aircraft and cause loss of control. The loss of pressurization from a door opening in flight also requires the use of oxygen by passengers at higher altitudes and loss of heat in the aircraft.

But the greatest danger when a door is lost concerns its effects on the control system of the aircraft. All wide-body jets are divided into a passenger cabin above and a cargo hold below (see fig. 3.1). Inside the floor that separates these two compartments are hydraulic lines, some control cables, and wiring that goes from the cockpit back to the aft engine and flying surfaces on the tail. Hydraulic pistons provide power to move the flying surfaces of jet aircraft, but

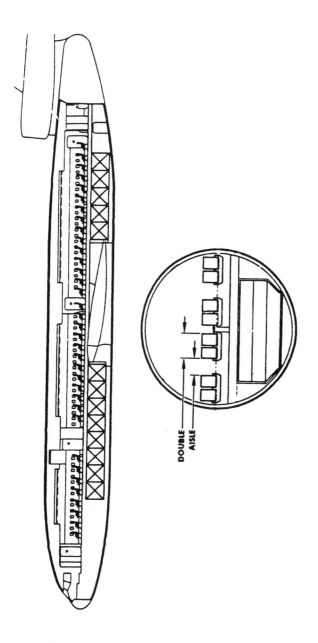

FIG. 3.1
DC-10 Floor Plan

they are activated by the use of control cables, which run from the cockpit to a hydraulic actuator. The Boeing 747 has control cables in the ceiling of the passenger cabin, but the Lockheed L-1011 and the DC-10 have their control cables in the floor along with the hydraulic lines. Damage to the floor is therefore a serious matter, for failure of the control cables or hydraulic lines will threaten the pilot's ability to control the aircraft.

Although the DC-10 will fly without the aft engine, loss of control of the horizontal stabilizers results in an extremely critical situation. These horizontal "wings" on the tail are used to control the "pitch" of the aircraft, its rotation around an axis passing along the length of the wings, i.e., the up-and-down movement of the nose and tail. The horizontal stabilizers on the tail are needed to counteract the nose-heavy design of commercial jets. Without their control, it is difficult to prevent the nose of the aircraft from pointing down. With a fully loaded aircraft, loss of tail control virtually guarantees that the plane will crash.

Wide-body jets have more than one set of hydraulic lines. The DC-10 has three, and the Boeing 747 and Lockheed L-1011 have four. These back-up systems ensure that failure of one hydraulic system will not disable the aircraft. Still, all hydraulic lines run through the floor, and in the DC-10 and L-1011 all control cables do so as well. A severe failure of the floor can endanger these aircraft despite the extra hydraulic systems.

The most likely cause of such a failure concerns the forces on the floor that result from pressurization of the aircraft. As altitude increases, air pressure falls, making it more difficult to breathe. In order to fly higher than about 5000 feet, aircraft must be artificially pressurized. Air is pumped into the plane to pressurize it; the higher the altitude the greater the pressure must be to create a comfortable environment for passengers and crew.

The additional force produced by pressurization is roughly five to ten pounds per square inch on the floor and walls of the aircraft. This is equivalent to an additional 720 to 1440 pounds on each square foot. The floors are able to withstand this additional load because both the passenger cabin and the cargo hold are pressurized. As a result, pressure above and below the floor is equal. Without this equalizing pressure the floor would not be able to withstand the force of pressurization. This means that if a door accidentally opens, the rapid depressurization (decompression) of the cargo hold or the passenger cabin would result in loss of counteracting pressure on the floor. Without counteracting pressure from the other side, the floor will collapse.

Thus there is an intimate relationship between the integrity of the doors and the ability to control the aircraft. Rapid decompression will collapse the floor and likely disable some, if not all, of the control cables, hydraulic lines, wiring, and fuel lines that run through the floor. A severe decompression accident would destroy all controls to the tail and result in loss of the aircraft. The safety of wide-body jets depends greatly on the integrity of their doors.

Doors

The design decision to place the control cables in the passenger cabin floor resulted in increased vulnerability of this type of aircraft to inadvertent opening of doors in flight. Each manufacturer took steps to prevent such accidents. Passenger cabin doors in all U.S. jumbo jets are plug or semi-plug doors. Because the door is larger than its opening (which it "plugs"), increased cabin pressure forces the door more tightly shut, making them virtually impossible to open in flight. Hence the danger from a passenger door opening are extremely remote.

Cargo doors are a different matter and reflect the design philosophies of the manufacturer. Lockheed installed semi-plug doors on the cargo hold of its L-1011. These doors incur a weight penalty, and require more extensive design, but they are extremely safe. The cargo doors on the Boeing family of wide-bodied jets also use an established and reliable latch system that protects against inadvertent opening in flight. The engineers at McDonnell Douglas developed a simple, over-center latch design which they believed would provide the same level of reliability and safety as the Boeing and Lockheed designs.

DC-10 Cargo Door Latch. The DC-10 cargo door is similar to the design of an automobile trunk lid. Closing the trunk is effected by a hook-shaped latch in the lid which grips a metal loop in the door frame. A lock then assures that the hook cannot open. The DC-10 cargo door is hinged from the top and opens like a car trunk. In the bottom of the door are a series of hooks (the latches), which attach to a corresponding series of short bars set in the bottom of the door frame (see fig. 3.2). The latches are mounted on a shaft which, when rotated by an electric or hydraulic actuator, hooks the latches around the bars (see fig. 3.3).

The latch system is designed so that when the mechanism goes over center (see fig. 3.4), forces on the latch caused by pressurization act to hold the latches in place. It is the same principle as an ordinary

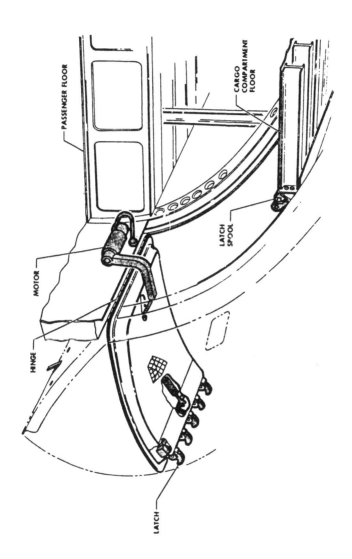

FIG. 3.2
DC-10 Rear Cargo Door

74

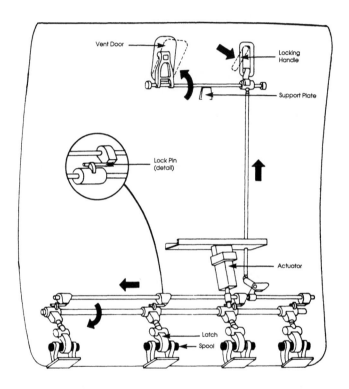

FIG. 3.3
DC-10 Rear Cargo Door Latch and Lock System

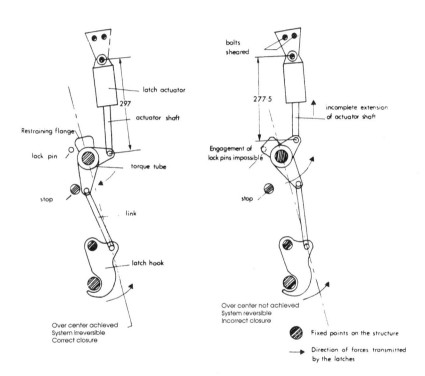

FIG. 3.4
DC-10 Latch Closing System

household light switch; once the switch bar goes past the center of its travel it moves quickly to the end position and resists being moved.

However if the actuator fails to rotate the shaft holding the latches far enough, over center is not achieved and pressurization forces will then be transmitted through the latches to the actuator (fig. 3.4). The actuator is not designed to withstand the forces produced when the latches are not over center, so in that situation it will fail, the latches will release, and the door will open. Recall that the 720 to 1440 pounds per square foot created by pressurization are also pressing on the door, trying to push it open.

DC-10 Cargo Door Lock. The cargo door lock mechanism is designed to perform two functions: to ensure that the latches are over center, and to prevent the latches from reversing. To achieve these ends, locking pins are mounted on a shaft parallel to the one holding the latches (see fig. 3.3). When the locking handle is manually depressed by a baggage handler, the locking pins move past corresponding flanges on the latches. As shown in figure 3.4, the pins can move past the flange only if the latches are over center, thus ensuring that the door is properly latched. And the pins having gone past the flanges, the latches cannot reverse because they will be stopped by the locking pins.

Latch Actuator. The actuator provides the force needed to rotate the latch shaft so that the latches curl around the fixed bars on the threshold of the cargo door. However, because of the vulnerability of the DC-10 floor, the choice between an electric or hydraulic actuator is critical. Each kind of actuator has a very different failure mode with the DC-10 latch system.

Consider the sequence of events that will transpire if, for some reason, the latches have failed to go over center and the door is not properly locked. Pressure against the door is transmitted up through the latch linkage to the actuator (see fig. 3.4). With an electric actuator, after it has rotated the latch shaft, a ratchet engages the actuator shaft and prevents it from moving. Any pressure on the actuator shaft is resisted by the internal ratchet and the two quarter-inch bolts holding the actuator in place. As the plane gains altitude, the pressure on the door increases. At some point the force on the actuator shaft will be great enough to cause the bolts holding it to fail. This point is at about 11,000 feet. Thus when the electric actuator fails, the pressure differential on the floor will be relatively high and its collapse a certainty. Thus the failure mode of the door with an electric actuator makes it more likely that loss of control will be substantial and that the aircraft will crash.

In contrast, a hydraulic actuator fails more gracefully. Pressure on the actuator shaft builds up as the plane gains altitude, but the hydraulic actuator is held in place by continuous hydraulic pressure (rather than a metal ratchet), so that the shaft will reverse when the force on the shaft exceeds the pressure in the hydraulic system. At that point, the actuator will reverse, opening the latches. What is important for the DC-10, is that it takes far less force to reverse the hydraulic actuator than to break the bolts that hold an electric actuator to the door. That means that the hydraulically actuated door will open at a much lower altitude where there is less internal pressure in the aircraft. This, in turn, means that the resultant pressure differential on the passenger cabin floor is much less, so that any buckling of the floor is less likely to severely damage control lines to the tail.

A hydraulic actuator would therefore act as a back-up to the lock mechanism. Inadvertent opening of an improperly latched and locked cargo door would not result in loss of the aircraft, because the failure would occur at an altitude which would not cause collapse of the floor and consequent loss of control. The DC-10 was originally designed to have hydraulic actuators for the cargo doors. They were replaced with electric actuators at the request of American Airlines in order to save on weight and maintenance costs. Hydraulic actuators are heavier than electric and they are more trouble to service. With this design change, the lock mechanism became the primary protection against improperly latched doors. Unfortunately, it had important weaknesses.

The Vent Door. During the initial ground testing in 1970 an improperly locked cargo door blew open and caused part of the floor to collapse (see "The 1970 Ground Testing Incident"). To increase the safety of the door, a vent door was added to the cargo doors. The vent door is a small plug door set into the cargo door (see fig. 3.3). It is connected to the lock mechanism, and its purpose is to prevent pressurization of the cargo cabin if the door has not been locked. The vent door closes when the locking handle is pushed home, so that failure to lock the door results in a visual indication of an unlocked door (the open vent door) and inability to pressurize the aircraft.

PROBLEMS WITH THE LOCK MECHANISM

The actual design of the locking mechanism neutralized its good features. There were two major defects of the design. First, the lock mechanism was not strong enough. It was possible to force the locking handle down with the locking pins jammed against the latch flange rather than passed behind them. Thus the door would appear to be locked, but in fact the latches would have failed to go over center and pressure on the door would force them open. The reason for this failure had to do with the weakness of the torque tube and the adjustable linkages in the lock mechanism (see fig. 3.5). An improperly adjusted lock mechanism would allow the locking handle to be closed even though the door was not fully latched.

Second, the vent door closure did not depend upon the movement of the locking pins at all, since it was on a different shaft (see fig. 3.3). In contrast, the vent door on the Boeing design (fig. 3.6) was located in the mechanism in such a way that it would close only if the locking pins had moved forward the correct distance (and were not jammed against the latch flange). Vent door closure on the DC-10 was *coincidental* to the movement of the locking pins, whereas in the Boeing design the vent door closes as a *consequence* of the movement of the locking pins.

The Windsor Incident

In 1972 the weakness of this latch and lock design was graphically demonstrated when an American Airlines DC-10 lost a rear cargo door and consequently suffered damage to the floor (see the "National Transportation Safety Board Report on the Windsor Incident"). The aircraft did not crash because of its light load (only sixty-seven passengers and crew) and the skill of its pilot, Bryce McCormick. A baggage handler had pushed the locking handle down by using his knee to force it closed.

After this incident,[1] a small viewing port, which allowed one to see the position of a lock pin, was placed in the door. In addition, the linkage was to be strengthened by the placement of a brace under the torque tube (fig. 3.3). Heavier wiring was required for the actuator to provide more power to move the latches to their over-center position.

The Paris Crash

The crash outside Paris in 1974 of a DC-10 flown by Turkish Airlines was caused by the failure of the rear cargo door at 11,500 feet. The passenger cabin floor collapsed, resulting in loss of control of the tail of the aircraft. The locking mechanism failed because (a) a support plate which was supposed to be added after a 1972 door failure

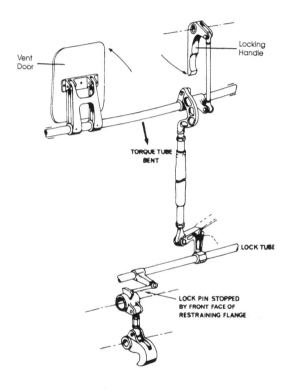

FIG. 3.5
DC-10 Rear Cargo Door Lock System

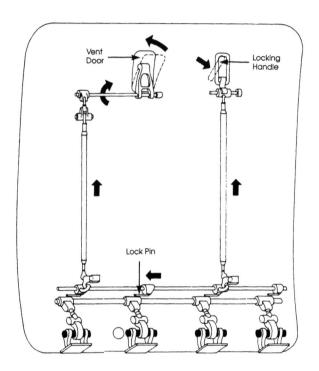

FIG. 3.6
Boeing-type Latch and Lock System

(see fig. 3.3) had not been installed despite records by McDonnell Douglas personnel certifying that the changes had been made; (b) the locking pins were not rigged properly, possibly by Turkish Airlines, making it easier for the locking handle to be closed even though the latches had not gone over center. As a result, the lock handle required only 50 pounds of force to depress instead of the 430 pounds that should have been required to push the handle down when the lock pins were jammed against the latch flanges. The lock mechanism could not serve to ensure that the latches were over center, and it was only a matter of time until the door failed.

A complex set of design decisions involving location of control systems, latches and locks, type of actuator, and how to modify a deficient design produced an accident waiting to happen. No one of those decisions was clearly wrong, but their combined effect was to produce a plane that was terribly vulnerable to decompression accidents and a cargo door whose weaknesses would cause it.

NOTE

1. In the language of aircraft accident reporting, an "incident" is a serious failure which does not result in a crash (an "accident").

PAUL EDDY
ELAINE POTTER
BRUCE PAGE

4

The 1970 Ground Testing Incident*

There were inherent weaknesses in the design of the DC-10 and, given the vulnerability of the cabin floor and its crucial role in carrying the flight controls, the cargo door locking system was quite inadequate. Often, such perceptions are available only in hindsight. But what the FAA seemingly did not know was that some well-qualified members of the DC-10 design teams were aware almost from the first that the cargo door system was suspect.

The detail design of the DC-10 fuselage and its doors was largely done by engineers from the Convair division of General Dynamics in San Diego, California. This was for financial as well as industrial reasons: McDonnell Douglas was naturally looking for subcontractors capable of sharing the enormous load of financing a program which could not run into profit for many years. (This is

---

* Reprinted, with editorial changes, from *Destination Disaster: From the Trimotor to the DC-10*, by Paul Eddy, Elaine Potter, and Bruce Page, Copyright © 1976 by Times Newspapers, Ltd. with permission of Times Books, a division of Random House, Inc.

standard procedure in the U.S. aerospace industry. Subcontractors pay their own start-up and engineering costs, getting the money back bit by bit over a predetermined period as they deliver units to the main contractor.)

Convair and its unsuccessful rivals for the contract (North American Rockwell, Rohr Aircraft, and Aerfer) were given, early in 1968, a weighty Subcontractor Bid Document which set out Douglas's requirements for the DC-10. The passenger doors were to be plugs, but the lower cargo doors were to be outward-hinging tension-latch doors. They were to have over-center latches driven by *hydraulic cylinders*, a system already used on some DC-8 and DC-9 doors. In addition to hydraulic latches, each cargo door was to have a manual locking system "designed so that the handle or latch lever cannot be stowed unless the door is properly closed and latched." The bid document was insistent that weight should be saved wherever possible (paint was to be used "to a minimum"), and the subcontractors were told that one pound of weight should be thought of as costing $100.

On August 7, 1968, McDonnell Douglas signed a subcontract with Convair. William Gross of Douglas later said that the reason for choosing Convair, in addition to the financial strength of the parent company, General Dynamics, was the excellent reputation Convair had for structural design. If that was so, it is surprising that Douglas did not take more notice of Convair's reservations about the DC-10 cargo door design. These began to emerge in November 1968, when Douglas told the San Diego engineers that instead of hydraulic cylinders they must use *electric* actuators to drive the cargo door latches.

Douglas gave two reasons for changing from hydraulic actuators: to do so would save twenty-eight pounds of weight per door and also would "conform to airline practice." The fact was that American Airlines had asked for electric actuators, saying that as they would have fewer moving parts they would be easier to maintain. And in the competitive conditions of the airbus market, requests of this kind from airlines were not likely to be rejected.

Some Convair engineers, and in particular their director of product engineering, F. D. "Dan" Applegate, were never fully reconciled to the change. The hydraulic system was, in Applegate's judgment at least, better because it was more "positive." When engineers say that one system is more positive than another, they are making a value judgment rather than a precise, mathematical statement just as they do when they say that one system is "simpler" than another. Nonetheless, the Convair argument has much substance to it.

Hydraulic latches would have been intrinsically safer with even a mediocre manual-lock system to back them up. But once the changeover was made to electric power, it became essential to provide a totally foolproof checking-and-locking back-up.

In the summer of 1969 Douglas asked Convair to start drafting a Failure Mode and Effects Analysis (FMEA) for the lower cargo door system of the DC-10. The purpose of the FMEA is, as the name suggests, to assess the likelihood of failure in a particular system, and the consequences of failure should it occur. Before an airplane can be certificated, the FAA must be given an FMEA for those major systems which are critical to safety.

Convair submitted a draft FMEA for the door system in August 1969. The design examined was an early one, in which the back-up locking system consisted simply of spring-loaded locking pins (later, of course, the spring-loaded pins were replaced by the manual locking handle and its complex linkages). Convair apologized for having taken two months over the work, due to the fact that their engineers had been working lately on military programs and were not familiar with current FMEA procedures for civil aircraft. In spite of this, and the fact that the door design analyzed was a relatively early one, Convair produced a document which accurately foresaw the deadly consequences of cargo door failure.

Among the "ground rules and assumptions" of the FMEA, those dealing with failure-warning systems stand up especially well to hindsight. First, said Convair, no great reliance was to be given to warning lights on the flight deck because "failures in the indicator circuit, which result in incorrect indication (i.e., 'lights out') of door locked and/or closed, may not be discovered during the checkout prior to take-off."

Convair claimed that even less reliance should be placed on warning systems which relied on the alertness of ground crews. In this early design, the only way of telling from the outside whether the latches had gone home was to look at the "manual override" handle provided to wind them shut by hand in case of electrical failure. If it had moved through its full travel, the latches must be safe. The Convair FMEA found:

> . . . That the ground crew requirement to visually check the angular position of the manual override handle, to detect an "unlocked" condition, to be subject to human error. It is assured that routine handling of repetitive aircraft could result in the omission of this check or visual error due to the location of the handle on a curved surface under the lower fuselage.

The substance of the FMEA then went on to show that there were nine possible failure sequences which could lead to a "Class IV hazard"—that is, a hazard involving danger to life. Five of these involved danger to ground crew by doors falling suddenly shut or coming open with undue violence, and these do not concern our narrative. But there were four sequences shown as capable of producing sudden depressurization in flight: also a "Class IV hazard," and meaning in this context the likely loss of the airplane. One of these sequences was, in principle, remarkably similar to what actually occurred over Windsor and later outside Paris.

The starting point was seen as a failure of the locking-pin system, due to the jamming of the locking tube or of one or more of the locking pins. In that case, said the FMEA: "Door will close and latch, but will not safely lock." There should of course be a warning against this, and if it works properly: "Indicator light [in the cockpit] will indicate door is unlocked and/or open." But one of the ground rules of the FMEA was that circuit failures in the indicator system might well go undetected, in which case: "Indicator light will indicate normal position." If that happened, malfunction of the electric latch actuators could produce a situation in which the "door will open in flight—resulting in sudden depressurization and possibly: structural failure of floor; also damage to empennage [tail] by expelled cargo and/or detached door. *Class IV hazard in flight.*"

The difference between this and what actually happened over Windsor is: first, the FMEA envisaged failure in a spring-loaded, rather than a hand-driven, locking-pins system; second, the failure mentioned was one of inadvertent electrical reversal of the latches rather than a failure of the latches to go over center in the first place. But it was a powerful demonstration that the door design was potentially dangerous without a totally reliable fail-safe locking system.

The other three depressurization sequences were rather different in that they envisaged total latch failure due to electrical faults, with the door being held shut until danger point by just the electric system which closes the door. These were not relevant to the accidents which ultimately occurred; nonetheless, together with the FMEA's general skepticism about warning lights and ground-crew assessments, they helped to produce a document that spelled out very clearly the terrible consequences that could follow from ill-thought-out door design.

But neither this FMEA draft nor anything seriously resembling it was shown to the FAA by Douglas, who, as lead manufacturer, made themself entirely responsible for certification of the airplane. (Indeed, under the terms of the subcontract, General Dynamics was forbidden from contacting the FAA about the DC-10). Our evidence is drawn from documents produced by Douglas and testimony given in the complex of compensation lawsuits which resulted from the Paris crash.[1] Evidence given by J. B. Hurt, Convair's DC-10 support program manager during the litigation, was that Douglas never replied to the Convair FMEA.

FMEAs submitted by Douglas to the FAA, leading up to certification of the DC-10, do not mention the possibility of Class IV hazards arising from malfunction of lower cargo doors.

The documentary warning of the dangers of depressurization were followed in 1970 by a physical manifestation at Long Beach. But even this it seems could not dent the self-assurance of the Douglas design team.

By May 1970, the first DC-10 (Ship 1) had been assembled at Long Beach and was going through ground tests to prepare for the maiden flight scheduled for August. On May 29, outside Building 54 the air-conditioning system was being tested, which involved building up a pressure-differential inside the hull of four to five pounds per square inch. Suddenly the forward lower cargo door blew open. Inside, a large section of the cabin floor collapsed into the hold.

The Douglas response to this foreshadowed the company's response to later and more serious accidents. It was simply blamed on the "human failure" of the mechanic who had closed the door. This explanation was still adhered to by William Gross when he gave evidence during the Paris crash litigation in 1974 and 1975. He gave no sign of thinking that there might be something basically dubious about a system that could become dangerous simply because one man, fairly low down in the engineering hierarchy, failed to perform exactly to plan.

However, Douglas did at the time acknowledge that some modification of the door was required before presenting the whole system for FAA certification. It had already been decided before the Ship One accident that the spring-loaded locking-pin system should be replaced by hand-driven linkages, and now it was decided to try and build some extra safeguards into that system. Ship One took off on its maiden flight on schedule with an unmodified door. But by the autumn of 1970 the "vent door" concept had been adumbrated. The miniature plug door was let into the main door above and to the right of the locking handle. This was supposed to stand conspicuously open until the main door was latched, and the locking handle was pulled down to drive the locking pins home.

In truth, it added little or nothing to the safety of the system. Such a vent door can only provide a check on the position of locking pins if its closure is a *consequence* of the pins having gone home. This is the case with the Boeing 747 tension-latch cargo door, which was already flying by 1970. But in the Douglas scheme, the closure of the vent door was merely *coincidental* to the action of the locking pins. If the rod transmitting the locking-handle's movement were to break, or to be absent altogether, then the Douglas vent door would still close.

On the face of it it seems especially remarkable that such an obvious flaw should have been overlooked. The certification process which the FAA imposes on every new airplane is supposed to identify and reject the offspring of dubious design philosophy. But the process, long and exhaustive thought it undoubtedly is, suffers from a fundamental weakness: Although it is carried out in the name of the FAA, much of the work involved is actually done by the manufacturers themselves.

The FAA says it has neither the manpower nor, in some instances, the specialized expertise to inspect every one of the thousands of parts and systems that go to make up a modern airliner. It therefore appoints at every plant designated engineering representatives (DERs)—company men, paid by the manufacturer, who spend part of their working lives wearing, as it were, an FAA hat.[2] Their job during the certification process is to carry out "conformity inspections" of the plane's bits and pieces to ensure that they comply with the Federal Airworthiness Regulations. In the case of the DC-10, there were 42,950 inspections. Only 11,055 were carried out by FAA personnel. The rest were done by McDonnell Douglas DERs.

Designated representatives are chosen by the FAA with a careful eye to their experience and integrity, but inevitably conflicts of interest can arise when manufacturers are called upon, in effect, to police themselves. The system also reduced the chance of mistakes being spotted, and the DC-10 vent door system stands as a classic illustration of what can happen. Before certification the vent door system was submitted to a series of tests by McDonnell Douglas and the results were approved, on behalf of the company, by a senior engineer. *Later, this time wearing his DER hat, the same engineer approved the report of the tests as acceptable documentation for showing that the DC-10 cargo door complied with the airworthiness regulations.*

There were other faults in the design. The linkages had not

been stressed correctly, so that they later turned out to be capable of flexing out of shape when submitted to pressure. And their various degrees of travel were all adjustable, so that the whole system was equivocal. Its validity as a check on the function of the latches depended upon whether the linkages in any given door happened to be correctly rigged.

Conceivably such design faults resulted from inexperience with doors of this kind. (Incidentally, if Douglas needed proof of the efficacy of the hydraulic system they had abandoned in the DC-10 cargo doors, it was abundantly available during 1970. There were five examples that year of hydraulic tension-latch doors in DC-8s and DC-9s opening in flight *before* pressure had built up to a dangerous level. All the aircraft landed safely.) But inexperience is no excuse, because in safety matters most airplane designers are willing to render assistance across competitive commercial boundaries.

Indeed, during November 1970, Convair was able to obtain via American Airlines considerable detail on Boeing vent systems for tension-latch doors. Not only did they discover that the two vent doors in the 747's cargo door were driven off the locking tube itself: they also learned that the locking system consisted of nonadjustable, and so unequivocal, linkages.

Not that everyone at Douglas's Long Beach plan thought that things were going the right way. In November a Convair engineer named H. B. ("Spud") Riggs, who was attached to the Douglas design team, wrote an internal memo headed: "Approaches to Eliminate Possibility of Cabin Pressurization with Door Unsafe." Riggs wrote that the design conception of the vent door was so far "less than desirable." He canvassed other ways of dealing with the problem: going back to hydraulic actuation; adding redundant electric circuits as back-up on the electric actuators; interlinking the door-closing system with the pressurization system; increasing the floor strength sufficiently to make the floor resist any possible pressure differential after a door blowout; providing vent space in the floor to enable high-pressure air to escape without doing damage.

None of these possibilities was incorporated in the DC-10—nor was Riggs's memo given to the FAA. By the following month Douglas as a corporation seemed to be more concerned about the financial consequences of the May blowout than the engineering ones. As with the doors, Convair was responsible for the detailed design of the DC-10 floor—albeit to Douglas's specifications. On December 4 an internal Convair memorandum recorded a negotiation with Douglas officials:

Douglas indicated that there was nothing defective in the door but that the passenger floor, having failed in the pressure test, was defective, and Convair owed Douglas a new floor. However, since it was not practical to change the floor Douglas wanted Convair to install the blow out door [vent door] in the 300 aircraft to satisfy its obligation on the floor.

There had always been a case for making the floor of the DC-10, and all other wide-bodied jets, strong enough to withstand full pressure differential. To do so would have cost some three thousand pounds in the DC-10 and the TriStar: say, one dozen passengers and their luggage, which in terms of present load factors might seem quite tolerable.

But that had been rejected—by Lockheed and Boeing, as well as Douglas—and the floor which Convair had built was by this time exactly as strong as Douglas had specified. Now Douglas was trying to say that the floor was not really strong enough ("defective"), so the door needed to be more reliable; therefore, the insertion of the vent door should be paid for by Convair.

On December 15, 1970, another Convair memo noted that Douglas had decided that in all cargo versions of the DC-10 the *upper* cargo doors would not have electric actuators and vents, but would go back to hydraulic actuation. M. R. Yale, Convair's manager of DC-10 engineering, wrote: "[We] asked Douglas why this approach would not be a better solution to lower cargo doors than vent doors. Only answer received was that Douglas had considered these factors and concluded that the vent door was the appropriate solution."

In the new year of 1971, with seven months to go before the scheduled date for the certification of the DC-10 as a commercial airliner, Douglas formally directed the installation of vent doors on all lower cargo doors, and although they were delayed by tooling problems and shortages of parts, deliveries were getting under way in June.

On July 6, 1971, a Convair memo gave a résumé of the situation:

(1) Design criteria and design features of operating, latching, and locking mechanisms were specified by Douglas for Convair.

(2) Basic design work was done by Convair engineers working at Long Beach under supervision of Douglas Engineering Department. Douglas retained total responsibility for

obtaining FAA approval of DC-10 and prohibited Convair from discussing any design feature with FAA.

(3) After the 1970 incident Douglas unilaterally directed incorporation of the vent door, even though there were in Convair's opinion several simpler, less costly alternative methods of making the failure more remote.

On July 29, 1971, the DC-10 was certificated by the FAA. Less than one year later came the blowout over Windsor, Ontario.

NOTES

1. *Hope v. McDonnell Douglas et al.*, Civ. No. 17631, Federal District Court, Los Angeles, California.

2. The FAA also appoints designated manufacturing inspection representatives (DMIRs) to assist the agency in monitoring production.

5

National Transportation Safety Board Report on the Windsor Incident*

SYNOPSIS

American Airlines Inc., Flight 96, a DC-10-10, N103AA, was a scheduled passenger flight from Los Angeles, California, to LaGuardia Airport, New York, with intermediate stops at Detroit.

Fuel, cargo, and passengers were loaded aboard the airplane at Detroit and its takeoff gross weight was computed to be 300,888 pounds, well under the maximum allowable. The last compartment to be secured prior to dispatch of the flight was the aft bulk-cargo compartment.

The ramp service agent who serviced the aft cargo compartment had difficulty closing the door. He stated that he closed the

---

* Reprinted, with editorial changes, from *National Transportation Safety Board Report, NTSB-AAR-73-2* (Windsor).

door electrically. He listened for the motor to stop running and then attempted to close the door handle. This handle is designed to close the small vent door which is located on the cargo door, to position a lockpin behind a cam on each of the four latches, and to open the circuit to the cockpit warning lights.

The agent could not close the handle with normal force, so he applied additional force with his knee. This caused the handle to stow properly, but the vent door was closed in a slightly cocked position. The agent brought this to the attention of a mechanic who gave his approval for release of the aircraft. According to the flight engineer, the cargo door warning light on his panel never illuminated during the taxi-out or at any time during the flight. The light was designed to illuminate when any cargo door is not properly secured for flight.

About 1925, while the airplane was at approximately 11,750 feet and climbing at 260 knots indicated airspeed (KIAS), the flight crew heard and felt a definite "thud." Simultaneously, dust and dirt flew up in their faces, the rudder pedals moved to the full left-rudder position, all three thrust levers moved back near the flight idle position, and the airplane yawed to the right. The captain reported that he lost his vision momentarily; he thought that a midair collision had occurred and that the windshield had been lost.

At the time of the occurrence, most of the flight attendants hard a loud noise, observed "fog" in the cabin and felt motion of the cabin air. The decompression of the cabin air through the aft cargo compartment door caused the cabin floor in the aft lounge area to fall downward and partially drop into the cargo compartment.

The captain declared an emergency, and Air Route Traffic Control cleared the flight back to Detroit via radar vectors. The airplane touched down 1900 feet down the runway and immediately started to veer to the right. As the aircraft veered further right, the first officer applied full reverse thrust to the left engine and brought the right engine out of reverse. This action provided directional control and the airplane paralleled the right side of the runway for 2800 feet before it began a gradual left turn back to the runway. The airplane came to rest approximately 8800 feet from the runway threshold. The nose and left main landing gear were on the runway surface, and the right main landing gear was off the runway surface.

The captain ordered the emergency evacuation alarm activated after the airplane came to rest. The evacuation slides were deployed and all passengers and crew used the slides.

Analysis

An American Airlines, Inc., McDonnell Douglas DC-10-10, was damaged substantially when the aft bulk-cargo compartment door separated from the aircraft in flight at approximately 11,750 feet mean sea level. The separation caused rapid decompression which, in turn, caused failure of the cabin floor over the bulk cargo compartment. The floor partially collapsed into the cargo compartment, disrupting various control cables which were routed through the floor beams to the rear engine and to the empennage [tail] control systems. [See Fig. 5.1]

The separated floor caused minor damage to the fuselage above the door opening and substantial damage to the leading edge of the left horizontal stabilizer. There were fifty-six passengers and a crew of eleven aboard the aircraft. Two stewardesses and nine passengers received minor injuries.

The National Transportation Safety Board determines that the probable cause of this accident was the improper engagement of the latching mechanism for the aft bulk-cargo compartment door during the preparation of the airplane for flight. The design characteristics of the door-latching mechanism permitted the door to be apparently closed, when, in fact, the latches were not fully engaged and the latch lockpins were not in place.

Structural damage to the door verified that the latches were not over center when the door opened in flight. The two fasteners which attached the door latch actuator to the door had both failed in shear. Forces transmitted back through the linkages from the door latches are the only means by which the actuator and supporting bracket could be loaded in flight. These latch loads are translated back through the actuator only if the actuator linkages are not over center.

A subsequent test of the door mechanism demonstrated that the door handle could actually be stowed without the lockpins in place if a force of 120 pounds was applied to the handle. Deflection of the mechanism allowed this to happen. The same deflection might have permitted the pilot indicator switch to make contact, which, in this system, prevents illumination of the cockpit warning light. Thus, the crew had no warning that the door mechanism was not functioning properly. Such a switch contact was also observed in the test conducted at the manufacturer's facility.

The increased pressure differential between the pressurized bulk cargo compartment and the outside atmosphere during the climbout loaded the latches, which eventually caused failure of the

96

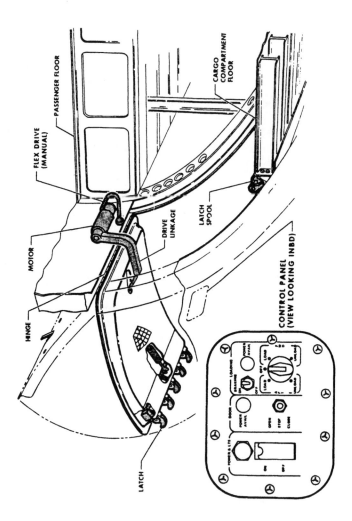

FIG. 5.1
DC-10 Rear Cargo Door

fasteners which secured the actuator support bracket to the door structure. The latches then sprung open, which permitted the door to blow open.

The loss of the aft cargo compartment door resulted in rapid loss of pressurization in that compartment. This particular cargo compartment was not equipped with pressure relief vents to the passenger cabin above it, as were the other cargo compartments of the airplane. Thus, the loss of the door caused the full differential pressure between the pressurized passenger cabin and the atmosphere to be exerted on the cabin floor above the compartment. This loading failed the floor support structure, and the cabin floor collapsed downward into the cargo compartment.

The collapse of the floor resulted in the loss of much of the control of the empennage control surfaces; although the airplane was designed with considerable redundancy in its flight control systems, the control cables from the cockpit to the empennage control actuators are routed through the cabin floor beams over this cargo compartment. The cabin floor displacement and floor beam deformation either severed or severely impaired the operation of these cables.

The board believes that the lack of pressure relief vents in this cargo compartment represents a significant hazard; sudden loss of pressurization in this compartment should not jeopardize the safety of the flight.

The crew reported extremely heavy control forces were necessary for pitch control. Two elevator control cables were separated and two remained intact; however, the downward loading of the floor on the cables made the use of increased forces necessary to move the control yoke.

UNITED STATES OF AMERICA
NATIONAL TRANSPORTATION SAFETY BOARD
WASHINGTON, D.C.

ISSUED: July 6, 1972

Adopted by the NATIONAL TRANSPORTATION SAFETY BOARD
at its office in Washington, D.C.
on the 23rd day of June 1972

FORWARDED TO:
Honorable John H. Shaffer
Administrator
Federal Aviation Administration
Washington, D.C. 20591

SAFETY RECOMMENDATIONS A-72-97 and 98

The National Transportation Safety Board is investigating an accident involving an American Airlines McDonnell Douglas DC-10-10, N103AA, which occurred shortly after takeoff from Detroit Metropolitan-Wayne County Airport on June 12, 1972.

The aft left-hand cargo door opened while the aircraft was at approximately 12,000 feet. The cabin floor over this cargo compartment then failed as a result of depressurization loading, and the floor dropped partially into the cargo compartment. This displacement of the floor caused serious disruption of the control cables which are routed through the floor beams to the empennage control systems and the engine controls. With the exception of the right rudder pedal cable, all of the cables on the left side of the fuselage broke. The cable runs on the right side were also damaged—the cable guides tore from their attachments to the floor beams, and the cables were deflected downward by the floor structure.

Preliminary investigation has revealed that the door latches were not driven fully closed, and the lock pins which should have prevented the latches from opening were not in place. The reason these door latches were not driven over center to the fully closed position has not yet been determined. However, although the Safety Board believes this was a relevant and contributing factor, we are more concerned over the failure of the door safety features to preclude dispatch of the aircraft with this door improperly closed. With the lockpins not engaged, a small vent door on the cargo door should have remained open, preventing normal pressurization of the aircraft. Also, the vent door handle should not have stowed

However, tests conducted at Douglas Aircraft Company have demonstrated that the vent door can be closed, and the handle stowed, without the lockpins engaged. Apparently, some combination of deflection of the operating mechanisms and tolerances permitted such operation when a force of approximately 120 pounds was applied to the operating handle. During these tests our investigator observed bending of the sliding lock-tube which caused the cap end of the tube to make contact with the pilot indicator switch actuating arm. This arm moves through a very small arc to activate the switch, and the Board believes that a similar contact on the accident aircraft door probably actuated the switch and gave the pilots a door safe indication on the annunciator light panel.

Finally, the Board believes that sudden loss of pressure in this cargo compartment for any reason should not jeopardize the safety of the flight. In this case, the loss of the door and resultant cabin floor failure caused an unwanted rudder input, severely restricted the elevator control available to the crew and disrupted the No. 2 engine controls.

We are aware of the inspection procedures currently in effect to ensure safety of operations of the DC-10 as well as the existing safety features of the floor design. Nevertheless, in order to preclude the recurrence of similar accidents, the Safety Board recommends that the Federal Aviation Administration:

1. Require a modification to the DC-10 cargo door locking system to make it physically impossible to position the external locking handle and vent door to their normal door locked positions unless the locking pins are fully engaged.
2. Require the installation of relief vents between the cabin and aft cargo compartment to minimize the pressure loading on the cabin flooring in the event of sudden depressurization of the cargo compartment.

Members of our Bureau of Aviation Safety will be available for consultation in the above matter if desired.

DEPARTMENT OF TRANSPORTATION
FEDERAL AVIATION ADMINISTRATION

WASHINGTON, D.C. 20590

7 JUL 1972

Honorable John H. Reed
Chairman, National Transportation Safety Board
Department of Transportation
Washington, D.C. 20591

Dear Mr. Chairman:

This is to acknowledge your Safety Recommendations A-72-97 and 98 issued on 6 July 1972 which included recommendations resulting from an inadvertent cargo door opening reported on a Douglas DC-10-10 airplane.

All operators of DC-10-10 airplanes are currently performing 100 hour functional checks on the cargo door system and will incorporate necessary modifications in accordance with McDonnell Douglas Service Bulletins 52-27 and A52-35 within 300 hours. These modifications pertain to improvements in the inspection and operation of locking and vent mechanisms.

Additional modifications to the cargo door locking and pressurization systems are being considered as part of a continued investigation effort. While a preliminary investigation indicates that it may not be feasible to provide complete venting between cabin and cargo compartments, your recommendations will be considered with respect to further action taken.

Sincerely,

J. H. Shaffer
Administrator

6

The Applegate Memorandum*

Fifteen days after Windsor, after the gentleman's agreement had been struck, and after Jack Shaffer had relaxed in the knowledge that Jack McGowen had fixed "his goddam airplane," Dan Applegate of Convair wrote a remarkable memorandum, which demands to be quoted in full. It expresses, with a vehemence not commonly found in engineering documents, all the doubts and fears that some of the Convair team felt about the airplane they were working on.

27 June 1972
Subject: DC-10 Future Accident Liability.

The potential for long-term Convair liability on the DC-10 has caused me increasing concern for several reasons.

---

* Reprinted, with editorial changes, from *Destination Disaster: From the Trimotor to the DC-10*, by Paul Eddy, Elaine Potter, and Bruce Page, Copyright © 1976 by Times Newspapers, Ltd. with permission of Times Books, a division of Random House, Inc.

1. The fundamental safety of the cargo door latching system has been progressively degraded since the program began in 1968.
2. The airplane demonstrated an inherent susceptibility to catastrophic failure when exposed to explosive decompression of the cargo compartment in 1970 ground tests.
3. Douglas has taken an increasingly "hard-line" with regards to the relative division of design responsibility between Douglas and Convair during change cost negotiations.
4. The growing "consumerism" environment indicates increasing Convair exposure to accident liability claims in the years ahead.

Let me explain my thoughts in more detail. At the beginning of the DC-10 program it was Douglas' declared intention to design the DC-10 cargo doors and door latch systems much like the DC-8s and -9s. Documentation in April 1968 said that they would be hydraulically operated. In October and November of 1968 they changed to electrical actuation which is fundamentally less positive.

At that time we discussed internally the wisdom of this change and recognized the degradation of safety. However, we also recognized that it was Douglas' prerogative to make such conceptual system design decisions whereas it was our responsibility as a sub-contractor to carry out the detail design within the framework of their decision. It never occurred to us at that point that Douglas would attempt to shift the responsibility for these kinds of conceptual system decisions to Convair as they appear to be now doing in our change negotiations, since we did not then nor at any later date have any voice in such decisions. The lines of authority and responsibility between Douglas and Convair engineering were clearly defined and understood by both of us at that time.

In July 1970 DC-10 Number Two[1] was being pressure-tested in the "hangar" by Douglas, on the second shift, without electrical power in the airplane. This meant that the electrically powered cargo door actuators and latch position warning switches were inoperative. The "green" second shift test crew manually cranked the latching system closed but failed to fully engage the latches on the forward door. they also failed to note that the external latch "lock" position indicator showed that the latches were not fully engaged. Subsequently, when the increasing cabin pressure reached about 3 psi (pounds per square inch) the forward door blew open. The resulting explosive decompression failed the cabin floor downward rendering tail controls, plumbing, wiring, etc. which passed through the floor, inoperative. This inherent failure mode is catastrophic, since it results in the loss of control of the horizontal and vertical tail and the aft center engine. We informally studied and discussed with Douglas alternative corrective actions including blow out panels in the cabin floor which would provide a predictable cabin floor failure mode which would accommodate the "explosive" loss of cargo compartment pressure without loss of tail surface and aft center engine control. It seemed to us then prudent that such a change was indicated since "Murphy's Law" being what it is, cargo doors will come open sometime during the twenty years of use ahead for the DC-10.

Douglas concurrently studied alternative corrective actions, inhouse, and made a unilateral decision to incorporate vent doors in the cargo doors. This "bandaid fix" not only failed to correct the inherent DC-10 catastrophic failure mode of cabin floor collapse, but the detail design of the vent door change further degraded the safety of the original door latch system by replacing the direct, short-coupled and stiff latch "lock" indicator system with a complex and relatively flexible linkage. (This change was accomplished entirely by Douglas with the exception of the assistance of one Convair engineer who was sent to Long Beach at their request to help their vent door system design team.)

This progressive degradation of the fundamental safety of the cargo door latch system since 1968 has exposed us to increasing liability claims. On June 12, 1972 in Detroit, the cargo door latch electrical actuator system in DC-10 number 5 failed to fully engage the latches of the left rear cargo door and the complex and relatively flexible latch "lock" system failed to make it impossible to close the vent door. When the door blew open before the DC-10 reached 12,000 feet altitude the cabin floor collapsed disabling most of the control to the tail surfaces and aft center engine. It is only chance that the airplane was not lost. Douglas has again studied alternative corrective actions and appears to be applying more "bandaids." So far they have directed us to install small one-inch diameter, transparent inspection windows through which you can view latch "lock pin" position, they are revising the rigging instructions to increase "lock pin" engagement and they plan to reinforce and stiffen the flexible linkage.

It might well be asked why not make the cargo door latch system really "fool-proof" and leave the cabin floor alone. Assuming it is possible to make the latch "fool-proof" this doesn't solve the fundamental deficiency in the airplane. A cargo compartment can experience explosive decompression from a number of causes such as: sabotage, mid-air collision, explosion of combustibles in the compartment and perhaps others, any one of which may result in damage which would not be fatal to the DC-10 were it not for the tendency of the cabin floor to collapse. The responsibility for primary damage from these kinds of causes would clearly not be our responsibility, however, we might very well be held responsible for the secondary damage, that is the floor collapse which could cause the loss of the aircraft. It might be asked why we did not originally detail design the cabin floor to withstand the loads of cargo compartment explosive decompression or design blowout panels in the cabin floors to fail in a safe and predictable way.

I can only say that our contract with Douglas provided that Douglas would furnish all design criteria and loads (which in fact they did) and that we would design to satisfy these design criteria and loads (which in fact we did).[2] There is nothing in our experience history which would have led us to expect that the DC-10 cabin floor would be inherently susceptible to catastrophic failure when exposed to explosive decompression of the cargo compartment, and I must presume that there is nothing in Douglas' experience history which would have led them to expect that the airplane would have this inherent characteristic or they would have provided for this in their loads and criteria which they furnished to us.

My only criticism of Douglas in this regard is that once this inherent weakness was demonstrated by the July 1970 test failure, they did not take immediate steps to correct it. It seems to be inevitable that, in the twenty years ahead of us, DC-10 cargo doors will come open and I would expect this to usually result in the loss of the airplane. [Emphasis added.] This fundamental failure mode has been discussed in the past and is being discussed again in the bowels of both the Douglas and Convair organizations. It appears however that Douglas is waiting and hoping for government direction or regulations in the hope of passing costs on to us or their customers.

If you can judge from Douglas' position during ongoing contract change negotiations they may feel that any liability incurred in the meantime for loss of life, property and equipment may be legally passed on to us.

It is recommended that overtures be made at the highest management level to persuade Douglas to immediately make a decision to incorporate changes in the DC-10 which will correct the fundamental cabin floor catastrophic failure mode. Correction will take a good bit of time, hopefully there is time before the National Transportation Safety Board (NTSB) or the FAA ground the airplane which would have disastrous effects upon sales and production both near and long term. This corrective action becomes more expensive than the cost of damages resulting from the loss of one plane load of people.

F. D. Applegate
Director of Product Engineering

Although this was not a formally set out safety analysis, like the Convair-drafted FMEA which Douglas did not give to the FAA, the Applegate memorandum was in some respects an even more disturbing document. Yet not only did its contents not reach the FAA, where surely they would have eroded even John Shaffer's durable complacency, they were not even put by Convair to Douglas.

The immediate fault in this must, of course, lie with Convair. But something must also be said about motivation: Briefly, Convair's experience over the previous three years had led to the belief that to raise such major safety questions with Douglas was chiefly to give away points in an ongoing financial contest. Certainly the record appeared to be one in which Douglas had been unimpressed by the safety propositions argued by Convair in its draft FMEA, and by the middle of 1972, it seems clear the position of the two companies was essentially an adversary one. Five days after Applegate wrote his memorandum his immediate superior, Mr. Hurt, wrote an equally revealing comment upon it:

3 July 1972
From: J. B. Hurt
Subject: DC-10 Future Accident Liability
Reference: F. D. Applegate's Memo, same subject, date 27 June 1972

I do not take issue with the facts or the concern expressed in the referenced memo. However, we should look at the "other side of the coin" in considering the subject. Other considerations include:

1. We did not take exception to the design philosophy established originally by Douglas and by not taking exception, we, in effect, agreed that a proper and safe philosophy was to incorporate inherent and proper safety and reliability in the cargo doors in lieu of designing the floor structure for decompression or providing pressure relief structure for decompressions or providing pressure relief provisions in the floor. The Reliance clause in our contract obligates us in essence to take exception to design philosophy that we know or feel is incorrect or improper and if we do not express such concern, we have in effect shared with Douglas the responsibility for the design philosophy.

2. In the opinion of our Engineering and FAA experts, this design philosophy and the cargo door structures and its original latch mechanism design satisfied FAA requirements and therefore the airplane was theoretically safe and certifiable.

3. In redesigning the cargo door latch mechanism as a result of the first "blowout" experience, Douglas unilaterally considered and rejected the installation of venting provisions in the floor in favor of a "safer" latch mechanism.[3] Convair engineers did discuss the possibility of floor relief provisions with Douglas shortly after the incident, but were told in effect, "We will decide and tell you what changes we feel are necessary and you are to await our directions on redesign." This same attitude is being applied by Douglas today and they are again making unilateral decisions on required corrections as a result of the AAL Detroit incident.[4]

4. We have been informally advised that while Douglas is making near-term corrections to the door mechanism, they are reconsidering the desirability of following-up with venting provisions in the floor.

I have considered recommending to Douglas Major Subcontracts the serious consideration of floor venting provisions based on the concern aptly described by the referenced memo, but have not because:

1. I am sure Douglas would immediately interpret such recommendation as a tacit admission on Convair's part that the

original concurrence by Convair of the design philosophy was in error and that therefore Convair was liable for all problems and corrections that have subsequently occurred.

2. Introducing such expression at this time while the negotiations of SECP 297 and discussion on its contractual justification are being conducted would introduce confusion and negate any progress that had been made by Convair in establishing a position on the subject.[5] I am not sure that discussion on this subject at the "highest management level" recommended by the referenced memo would produce a different reaction from the one anticipated above. We have an interesting legal and moral problem, and I feel that any direct conversation on this subject with Douglas should be based on the assumption that as a result Convair may subsequently find itself in a position where it must assume all or a significant portion of the costs that are involved.

J. B. Hurt
Program Manager, DC-10 Support Program

On July 5, 1972, Mr. M. C. Curtis, the Convair vice-president who was in overall charge of the DC-10 project, called a meeting to decide corporate policy in the light of the Applegate memorandum. (Hurt's memo was chiefly a briefing to Curtis for the meeting.) Convair's chief counsel, director of operations and director of contracts, attended along with Applegate and Hurt. (It seems that one person not consulted was David Lewis, the one-time heir apparent of "Miser Mac" who had become president of General Dynamics in 1970. Ironically, it was Lewis as president of the Douglas division who, in 1968, negotiated the financial details of the contract with Convair.)

It was acknowledged that Applegate was closer than Hurt to the engineering of the DC-10 and had a better knowledge of the safety factors involved. But Mr. Curtis and his colleagues preferred the reasoning of Hurt's memo and resolved the "interesting legal and moral problem" by deciding that Convair must not risk an approach to Douglas. According to Hurt's testimony, two-and-a-half years later in *Hope v. McDonnell Douglas*, the meeting came up with a rationalization which, though touched with cynicism, had justification of a sort. "After all," said Hurt, "most of the statements made by Applegate were considered to be well-known to Douglas and there was nothing new in them that was not known to Douglas." And it is certainly hard to believe that the Douglas design team could have claimed, after three years of close collaboration with Convair, that the arguments of the Applegate memorandum were unknown to them.

Both to Douglas and to Convair the dangers were, or should have been, obvious. The determination of Convair, as a subcontractor, was not to take upon itself the duty of pointing them out.

And so, because of the interrelated failures of McDonnell Douglas, Convair, and the Federal Aviation Administration, a fundamentally defective airplane continued on its way through the "stream of commerce" (as plaintiff lawyers like to call it).

NOTES

1. We have been unable to establish whether Applegate's reference to an accident involving Ship Two, in July 1970, is a mistake on his part of whether there were *two* blowout incidents. Certainly Ship One was damaged on May 29, 1970, in circumstances very similar to those described by Applegate.

2. Douglas's design criteria called for the floor to withstand a pressure of 3 psi, and it eventually did so—although not until Douglas had challenged Convair's original stress analysis, and a stronger kind of aluminum alloy had been introduced to the floor beams.

3. Vents built into the floor would allow pressurized air to escape into the hold without buckling the floor.

4. American Airlines. Detroit is where the DC-10 landed after the Windsor incident.

5. A reference to negotiations over the cost of fitting three vent doors in each of 300 projected aircraft: a total cost of $3 million.

PAUL EDDY
ELAINE POTTER
BRUCE PAGE

7

Fat, Dumb, and Happy:
The Failure of the FAA*

The Federal Aviation Administration, which regulates the aviation industry in the United States, has eleven regional offices scattered throughout the country, each one responsible for watching over the manufacturers, airlines, and airports operating in its area. The Western Region office of the FAA is responsible for California which contains the greatest concentration of airplane-building ability in the world. Besides Douglas, Lockheed, Convair, and North American Rockwell, there is a vast supporting cast of smaller firms, providing everything from airborne lavatories to navigation avionics.

In 1972 the head of the Western Region was a career public servant named Arvin O. Basnight. His office on Aviation Boulevard,

---

* Reprinted, with editorial changes, from *Destination Disaster: From the Trimotor to the DC-10*, by Paul Eddy, Elaine Potter, and Bruce Page, Copyright © 1976 by Times Newspapers, Ltd. with permission of Times Books, a division of Random House, Inc.

not far from Los Angeles International Airport, had on one of its walls a large picture of B-17 bombers in combat—a reminder that Basnight's early flying experience was gained in harsh circumstances during the great daylight air battles at the climax of World War II. Basnight projected an affable, relaxed manner, but it did not quite conceal the watchful gaze and studied speech of the experienced bureaucrat.

The FAA is typical of U.S. public services in that its professional administrators know that the titular head of the agency will usually be a political appointee, nominated by the White House. Sometimes these imported chieftains excel the professionals; sometimes they are incompetent at running the agency they are chosen for; and sometimes, they are a capricious mixture of the two. For professional public servants like Basnight the principle of survival, naturally, is to get through the bad periods with as little personal damage as possible. It is probably fair to describe Basnight as a man who applied the tactics he learned in wartime to the conditions of public service life. B-17 crews were trained to keep close formation, eschew individual action, and make sure that they had covering fire at all times.

There cannot be much doubt that at the time when Mr. Basnight was trying to cope with the aftermath of the Windsor incident of June 12, the FAA was going through a bad period—one from which it has not fully recovered even now. Like many other federal agencies at that time, it was suffering from the fact that the White House was inhabited by a group of men who meant to dismantle the complex counterbalances of American government and put in their place the personal rule of President Richard Nixon.

The revelations of criminal and covert behavior overshadow, in memory, the fact that the Watergate era brought at the same time an overt attack on the power and traditions of the federal civil service. Some of the energy of this attack derived from Republican sentiments which were legitimate and traditional, however much some people might disagree with them. Republicans dislike "big government" on ideological grounds. The great federal agencies are often cumbersome and expensive in operation, and are largely staffed by Democrats who sometimes fail to serve Republican administrations—and ends—with undivided enthusiasm. In the case of Nixon and his White House staff, there was no doubt something more—the authoritarian's instinctive dislike of a system which diffuses administrative power among many different centers. In any event, the Nixon White House, beginning in 1969, set out to make federal agencies "more responsive" to the president's will. The result, in terms of our particular story, was the crippling of the FAA and of the other federal agency chiefly concerned with aviation safety, the National Transportation Safety Board.

The FAA owes its decline, in large measure, to President Nixon's first choice of an administrator. In 1969 the president nominated—and Congress approved—John Hixon Shaffer, then fifty, an ex-Air Force officer with rock-solid Republican loyalties, but with very little experience of commercial aircraft to qualify him to head the most powerful and influential aviation-regulating body in the world.

In law, the FAA has great powers over the American aviation industry. It regulates, and must approve, every stage of the design and manufacture of every aircraft which is built in the United States. It can inspect manufacturing plants at will and certificates as airworthy each aircraft that is built. If the FAA decides, for any reason, to withhold certification, then the particular aircraft may not leave the ground even to remove itself from American airspace. The FAA also regulates and licenses the pilots that fly the airplanes and most of the airports they use and operates the air traffic control systems that rule over the airways. The agency does not formally investigate airplane accidents because that is the job of the NTSB. But if an NTSB investigation reveals the need for change in the design of an aircraft, or the way in which it is operated, the FAA can order the change—no matter how expensive or awkward—by issuing an airworthiness directive, which has all the force of federal law.

All airplane manufacturers, of course, modify their aircraft in the light of service experience. In practice, this means a flow of bulletins going out to all known users of the airplane in question, with directions on how to accomplish the recommended change. Where necessary, kits of parts will be sent out to the operator, though major reworking may require that the plane go back to the factory. The majority of these service changes are concerned with operating convenience or economy, rather than safety. They are therefore optional, and although the manufacturer may provide kits for little or no charge, the onus of cost remains formally with the operator.

An airworthiness directive from the FAA is quite different. There is nothing optional about it, because if it is not complied with, it becomes illegal to fly the airplane. As the name implies, an AD is only issued when airworthiness—and therefore safety—is directly concerned.

In practice, ADs do not normally "ground" airplanes, or even interrupt operations very drastically. They merely say that certain changes must be accomplished at particular moments in the airplane's regular maintenance cycle. But they are public documents, which the manufacturer's own service bulletins are not, and they are circulated automatically to the news media, to the aviation attachés of all foreign governments, and to the FAA's overseas offices. Naturally, they place the onus of cost firmly on the manufacturer, because the contract of sale always says that what the manufacturer agrees to supply is an *airworthy* machine. Airworthiness is defined by the existence of FAA certificates, and the issuing of an AD makes those certificates invalid if the directive is ignored.

There is and always has been a potential weakness in the constitution of the FAA. By act of Congress its responsibility is not only to regulate the safety standards of American aviation but also to promote its commercial success. It probably did not occur to very many people when the FAA was established in 1958 that there was likely to be any serious conflict between the two aims. After all, it must have seemed obvious that mass air travel could not establish itself commercially unless it gained a reputation for safety. And, equally obviously, no aircraft manufacturer's long-term interests can be served by building airplanes with anything less than the greatest possible safety standards.

Less obvious, perhaps, is the problem that the strains of commercial competition may be so acute that the men and organizations subjected to them can cease to be accurate judges even of their own interests, let alone the interests of other people. In theory, of course, it is for just such occasions that the FAA, in its role as the industry's policeman, exists. But the agency can only work effectively if the man appointed to head it can cope with the schizoid nature (part policeman, part promoter) of the administrator's role. He must be able to identify with the industry's viewpoint, while standing slightly apart. The trouble with John Shaffer was that he identified totally with the industry. Indeed, he was an embodiment of it.

Shaffer served with the U.S. Air Force from 1946 to 1953 and spent most of that time working with aircraft manufacturers developing new bombers such as the Boeing B-47. When he resigned from the Air Force, with the rank of lieutenant colonel, he deliberately took a job outside the aviation industry with the Ford Motor Company to avoid, he says, accusations that he had been "feathering my own nest." But after three years with Ford, he deemed it proper, in his own words, "to go home" and he became an executive—and, eventually, a vice-president and small stockholder—of TRW, Inc., one of the largest engineering subcontractors in the aerospace industry.

Basnight and his men had no choice but to be fairly tough in order to get cooperation from McDonnell Douglas when they started to look into the cargo door problem. On June 13, the day after the Windsor incident, FAA engineers contacted the Douglas plant at Long Beach to find out if there had been any previous problems with DC-10 cargo doors, or if its failure over Windsor had come quite out of the blue. The company omitted to hand over operating reports filed by the airlines using DC-10s and acknowledged only that there had been a few "minor problems." (Every untoward incident in an airplane's life is reported to the manufacturer, and the airlines have an obvious legal incentive to make their reporting systems accurate. But it is a weakness of the regulatory system that it is left to the manufacturer to decide whether to draw the FAA's attention to any particular report.)

Richard (Dick) Sliff, then head of aircraft engineering under Basnight, is a highly experienced test pilot, well acquainted with the fact that airplane systems nearly always give some kind of warning before they fail. He was "disturbed" by the company's attitude and "raised a fuss" to get the airlines' reports.

On examining the records, Sliff found that during the ten months of DC-10 service, there had been approximately one hundred reports of doors failing to close properly and that Douglas had already had to recommend modifications to the system. The trouble was that the electric actuators were not always succeeding in driving the over-center latches fully over the latch spools. They were sticking partway, requiring extra applications of electric power, and in some cases hand winding. (All DC-10 doors can be opened and closed by a hand crank, in case of power failure.) The Douglas engineers had proposed lubrication of the latch spools when the problem was first reported, but that had not been effective. They had then sent out a service bulletin recommending that the power supply to the electric actuators be rewired in a heavier gauge of wire, thus lessening transmission resistance and increasing the power developed by the actuators. The four operating airlines, United, American, National, and Continental, were still rewiring doors when the Windsor incident occurred, and on Captain McCormick's plane it had not been accomplished.

The Western Division office had certificated the DC-10 (on July 29, 1971), and among many other things that had meant certificating the cargo and passenger doors. Basnight, Sliff, and their colleagues were therefore familiar with the arguments about tension-latch doors and plug doors, and with the choice between C-latch and over-center locking systems. They were also aware, in general terms, of the relationship between door sealing, floor strength, and airworthiness in a plane which, like the DC-10, carries its control cables on the underside of the floor beams.

The NTSB team investigating the Windsor incident were suggesting that the DC-10 cargo door should be modified to make it "physically impossible" for the door to be improperly closed. And the NTSB also recommended that the cabin floor should be modified and strengthened to prevent its collapsing after a sudden decompression. (There was informal contact between NTSB and FAA headquarters throughout June. The NTSB made its formal recommendations to the FAA in writing on July 6.)

But in the immediate aftermath of the Windsor incident, neither the FAA's Western Division office nor anyone at the Douglas plant could see any neat and immediate engineering answer to the problem. And there were DC-10s taking off virtually every hour.

If the plane was not to be grounded—a step with serious economic consequences in the middle of the summer air-travel season—there would have to be an interim "fix": something which could be agreed upon and installed rapidly and which would ameliorate the situation until a proper redesign could be accomplished.

The problem was simple enough to define: Because of the inadequacy of the manual-locking mechanism, a man closing the cargo door could not be sure that the locking pins had actually gone home. Why not, therefore, place a small peephole made of toughened glass in the middle of the metal door skin over one locking pin? Then the man closing the door would be able to *see* whether the pins were safely home. Nobody in the Western Region office thought this was a complete solution, for the door sill of a DC-10 stands some fifteen feet above the ground, and each locking pin is less than two inches long. To make a proper inspection, each baggage handler would have to wait for the door to come down, and then move his mobile platform along to peer into the one-inch peephole. At night, he would need a flashlight, and it might in any case be difficult to see through glass streaked with oil, dirt, and water. Still, given that all electric actuators were rewired, and given suitable alerting of ground crews, this kind of change would make the DC-10 reasonably safe until something better could be worked out.

The one thing that no one in the Western Region office doubted was that the reworking of the door would have to be enforced by a series of airworthiness directives, and by Wednesday, June 15, a certain amount of drafting had been done. On the morning of June 16, four days after McCormick's landing at Detroit, a preliminary text of the first proposed AD was sent by telecopier to the FAA headquarters in Washington, although it did not go beyond giving mandatory force to the wiring changes that Douglas had already recommended. During the day Everitt Pittman of the Western Region office, guided by Sliff and consulting with engineers at the Long Beach plant, worked on a more elaborate text which would specify the size and location of the inspection hole and require warning words and diagrams to be put up alongside them. But although the day had started with the assumption that an AD would be issued against the DC-10, within a few hours, Basnight's staff began to realize that something had gone wrong.

Basnight had started his day early, for the case of the DC-10 door was a considerable crisis on his beat. At 8:50, he received a phone call from Jackson McGowen, president of the Douglas division of McDonnell Douglas. Late the night before, said McGowen, he had spoken on the phone with Jack Shaffer, Basnight's boss, in Washington. They had "reviewed" the work Douglas had done on beefing up the wiring in the DC-10 door system. The administrator, said McGowen, had been pleased to hear that this work had been going on in cooperation with his own FAA officials—so much so, it appeared, that he saw no need for any airworthiness directives to follow from the Windsor incident. The call to Basnight was clearly to inform him that there was an understanding between his boss, John Shaffer, and the builders of the DC-10. As Basnight received the message: "Mr. Shaffer . . . had told Mr. McGowan [*sic*] that the corrective measures could be undertaken as a product of a Gentleman's Agreement thereby not requiring the issuance of an FAA Airworthiness Directive."

It is hardly surprising that after this crushing defeat the Western office made no further attempts to issue ADs relating to the DC-10 door system. However, Sliff and his colleagues remained in touch with the engineers at the Long Beach plant, who did at least make some effort to honor the gentleman's agreement that Jackson McGowen had made on their behalf.

It was, of course, obvious to any serious engineer that the situation could not be left as it stood on Friday, June 17. In addition to the actuator rewiring, Douglas proposed three more changes to the door, all of which were sent out to the airlines by service bulletins from Douglas over the next two months.

First, in SB 52-35 they adopted the peephole idea. This was an "Alert" service bulletin, printed on blue paper, rather than white, to show airlines that it concerned a safety problem. Alongside each door frame there was also to be a decal showing diagramatically what a baggage handler would see if he looked in after the locking pin behind the peephole was safely home. This was to be labeled SAFE. There was another decal diagram, giving an idea what might be seen if the pin had for some reason not moved, and this was to be labeled UNSAFE. As Chuck Miller was to point out later, this would demand remarkable devotion from a baggage handler, working on a snow-swept airport in a January night.

Then, in SB 52-37 they went somewhat further, producing for the first time an approximation of a satisfactory design. Tests after Windsor had shown that the essential fault lay in the weakness of the linkages which were intended to drive the locking pins home. When the pins, on that occasion, encountered unclosed latches, then the locking handle outside the door ought to have become immovable in the ground crewman's hand. Instead, when he pressed down on the handle with his knee, the mechanism merely gave way, enabling him to push the handle right down into its slot. The top of the handle passed through the door skin and was attached to a "torque tube" just inside. This was supposed to revolve with the downward movement of the handle, thereby moving a crank, which in turn would push a long vertical rod going down to the latch assemblies in the lower door. There, the movement of another crank was to translate the vertical thrust into a horizontal sliding movement, and thus drive the bar with all four lock pins mounted on it.

But the torque tube was made to revolve in between bearings which were a long way apart, and the span was too much for its strength. In the Windsor case, the tube had simply bent out of shape when the other parts of the linkage refused to move. Unhappily, though, it did revolve a little—just enough to operate the other linkage working off it, which ran upward to shut the vent door, thus completing the ground crew's illusion that all inside was well (see fig. 3.5).

In SB 52-37 the Douglas engineers therefore recommended the fitting of a "support plate" to hold up the torque tube just beside the handle. It was simply an aluminum bracket fixed to the inner skin of the door, with a half-circle cut out in which the torque tube could rest and revolve. This, it was calculated, would prevent the tube sagging under any conceivable manual pressure which might be applied to the handle. In the same bulletin, they recommended that the linkages be adjusted so as to extend the normal travel of the locking pins by one-quarter of an inch. This would make the mechanism jam all the more conspicuously if unclosed latches were to stand in the path of the locking pins.

It was not a perfect solution. But if these changes had been properly carried out on every DC-10 built and put into service, then the slaughter of March 3, 1974, might well have been averted. The trouble, of course, is that they were not properly carried out. (SB 52-37 was sent out as a *routine* service bulletin, printed on white paper with no indication that it was vital to the DC-10's safety. As a result, very few DC-10s were modified with any alacrity. When the bulletin was issued on July 2, 1972, there were thirty-nine DC-10s in service, all of them operated by U.S. domestic airlines. Only five airplanes were modified within ninety days; eighteen were not modified until 1973; and one DC-10, owned by National, was still flying around without a support plate on March 5, 1974—nineteen months after the bulletin was issued.)

Because history is an unrepeatable experiment, we cannot prove that the extra urgency, legal weight, and publicity which go with airworthiness directives would necessarily have made the difference. But the crucial point is not so much the issuing or nonissuing of any particular directive, as the general determination on John Shaffer's part that the Douglas Company itself could be left to handle the matter in its own way. And, of course, the very way that the thing was done weakened the authority of Arvin Basnight and his staff. And the secrecy in which the whole business was accomplished was damaging also: Douglas employees later testified that they were simply unaware of the significance of the various things that were supposed to be done to DC-10 doors.

According to Shaffer, the term gentleman's agreement is an unhappy one, conveying the incorrect impression that his transaction with Jackson McGowen amounted to "a handshake behind the barn." Yet the fact is that at the time he agreed with McGowen that no AD was necessary, it would have been quite impossible for McGowen to have given him any detailed account of what might be done to make the door safe—for no one at the Douglas plant or in the FAA had yet worked out a detailed solution. Shaffer covers this point by saying, simply, that "you have got to have faith in people." But Shaffer should have been aware that the summer of 1972 was a moment when nobody in the Douglas division of McDonnell Douglas could be expected to give an objective judgment on any matter which might involve publicity for actual or potential drawback in the DC-10.

Briefly, it was a deeply inconvenient moment for Douglas to have an AD issued against their plane.

In 1971, it had seemed that Lockheed and the TriStar were to be destroyed by bankruptcy. Against all the odds, and against the ideological prejudices of both the American and British governments, the TriStar and the RB-211 had survived, with the result that the Lockheed sales team had returned to the field with a new and desperate vigor. The competition had reached the point where the allegiances of most major American airlines had been declared, and the vital need was to capture foreign airlines. Sales teams from both Lockheed and McDonnell Douglas had the same philosophy: that what mattered in 1972 more than actual numbers sold was to obtain first orders from certain airlines thought to occupy influential positions. Both Lockheed and McDonnell Douglas were now looking hard at Turkish Airlines, believed to be ripe for a purchase of new equipment and thought to be capable of influencing other Middle Eastern airlines by its example.

The two airbus makers had each set up round-the-world sales odysseys for summer 1972, using in each case a specially prepared airplane which would be laden with executives, engineers, salesmen, and publicity experts of all kinds. Jackson McGowen was proposing to lead the Douglas excursion in person, and the Windsor incident occurred just before the scheduled departure of "Friendship '72" from Long Beach.

The special unpleasantness of the kind of AD that Basnight's men began preparing after the Windsor incident was that it was not unlikely to be clear-cut and swiftly disposable. It could not be dismissed as a once-only problem that had been examined and cured—at least, not within the time-scale of Friendship's tour. Because ADs are distributed through the diplomatic network, the problem of the exploding door might well have pursued Friendship '72 the whole way round-the-world. (There is little doubt that the Lockheed team, on their own circumnavigation, would have found discreet ways of capitalizing on the issuing of ADs against the DC-10.)

Shaffer was replaced as FAA administrator by Alexander Butterfield, then a little-known aide to Bob Haldeman who had been in charge of internal security at the White House.

Butterfield was not a great success as FAA administrator and in February 1975 he was forced to resign by the Ford administration. But to his credit he did, two days after the Paris crash, order the establishment of an FAA ad hoc committee to investigate the history of the DC-10 and the role that the agency had played in its development.

Much of the report the ad hoc committee produced on April 19,

1974, was highly technical, but it amounted to a damning indictment of the DC-10 door design and of the regulatory system in the United States that had permitted its certification. On the subject of the DC-10 door, the report said:

> While there is no longer any doubt that [the DC-10 cargo door] is safe, it is an inelegant design worthy of Rube Goldberg. . . . The DC-10 intercompartment structures [partitions, floors, bulkheads] were not designed to cope with or prevent failures that would interfere with the continued safe flight and landing after the sudden release of pressure in any compartment due to the opening of a cargo door. It was therefore incumbent on McDonnell Douglas to show that loss of the cargo door . . . was "extremely remote" for compliance with Federal Airworthiness Regulation 25. . . . It would appear, in the light of the two accidents, that the level of protection and reliability provided in the cargo door latching, safety locking mechanisms, and the associated warning systems was insufficient to satisfy the requirements of FAR 25. Additionally, it now appears that the possibilities of improper door operation were not given adequate consideration for compliance with FAR 25.

On the subject of the gentleman's agreement the FAA report had this to say:

> Review of the FAA's . . . correction programs associated with the DC-10 airplane has again pointed out that the agency has been lax in taking appropriate Airworthiness Directive action where the need for ADs are clearly indicated. This situation is by no means unique to the DC-10 airplane or to the Western Region. So-called "voluntary compliance" programs have become commonplace on a large number of aeronautical products and in most, if not all, of the [FAA] regions. Voluntary compliance programs have been . . . most commonly implemented on the smaller private and executive aircraft types where the number of ADs issued is most likely to be considered a sales deterrent by the manufacturer.
> Many complaints on this issue have been received from a number of foreign airworthiness authorities who have airworthiness responsibility over U.S. manufactured products in their respective countries; and, for the most part, the FAA has ignored those complaints.

Dealing specifically with the Windsor incident, the report said:

> The agency was not effective in attaining adequate fleet-wide corrective action on a timely basis after problem areas were clearly indicated by the . . . accident. Non-regulatory procedures and agreements were used in lieu of established regulatory AD procedure [and] in the long run proved to be ineffective in correcting design deficiencies . . . to prevent reoccurrence of the [Windsor] accident.

After reading that report and conducting its own review of the FAA and the DC-10 affair, a special subcommittee of the House of Representatives concluded that between June 1972 and the Paris crash of March 1974, "through regulatory nonfeasance, thousands of lives were unjustifiably put at risk." In defense of the technical experts of the FAA it has to be said that in the thirteen months leading up to the Paris crash they did begin to display some unease about the fundamental hazard (the vulnerability of the DC-10's floor) that the Windsor incident had publicly revealed.

By February 1973 the technical experts at the FAA's headquarters in Washington, D.C. had decided there was a need to do something about jumbo-jet floors. In its letter, dated February 2, the FAA asked the manufacturers to consider two alternative proposals: (1) the vital control systems could be rerouted away from the floor; or (2) the floor itself could be strengthened by reinforcing the supports and by adding vents which, in the event of decompression would allow pressurized air in the passenger cabin to escape before its weight could collapse the floor.

Boeing was plainly displeased at being involved in the aftermath of Windsor. As the company's reply to the FAA pointed out, the 747's control cables do not run under the floor (although the hydraulic tubes do) and added: "In the decompression cases quoted in your letter, the door locking systems [on the DC-10] were not fail safe in that single failures permitted the pressurization of unlocked doors which led to the opening of the doors in flight."

Lockheed felt equally strongly that it was unfair to include the TriStar in any redesign program that the FAA obviously now contemplated because of Windsor. The company felt there was no question of TriStar doors coming open in flight because they were semiplugs. As for the danger of small bomb explosions (an alternative hazard to floors which the FAA had also raised in its letter), Lockheed argued that jumbo jets were no more vulnerable to this threat than all other pressurized airliners.

For its part, McDonnell Douglas displayed no embarrassment over the fact that it was the DC-10 that had caused all the trouble, insisting that the chances of a DC-10 cargo door opening in flight were "extremely remote." The company's reply to the FAA, dated March 15, 1973, continued: "A reassessment of the DC-10 design, with regard to the effects on safety, for nonplug cargo doors and small bomb explosions shows that the present standard and levels of substantiation [of the floor] are adequate."

It was not until February 5, 1974, that FAA headquarters formally repeated its request to the Western Region that McDonnell Douglas and Lockheed should be asked to carry out a technical study of jumbo-jet floors. This time the Western Region did at least pass on the request. Lockheed replied that the study called for would be expensive and time consuming and might be better done by the aviation industry as a whole, rather than by individual manufacturers. McDonnell Douglas said much the same. "We do not have the manpower available at this time to undertake this study, nor are we in a position to accept this burden alone."

In its reply to the FAA, dated February 25, 1974, McDonnell Douglas also said that if the government wanted a study made, then government funds should pay for it. Six days later came the Paris crash.

The irresistible question is: Could the Paris tragedy have been averted if the three manufacturers—and specifically McDonnell Douglas—had taken more notice of the concern over jumbo-jet floors? By February 1974 it was, of course, far too late for a technical study to have altered the fate of Turkish Airlines' DC-10. But in April 1971, when the Dutch RLD [equivalent to our FAA] first voiced concern, the DC-10 had not yet been certificated and only a handful of models had been built. True, it would have been a highly embarrassing moment—from a commercial viewpoint—to return to the drawing board, as it were, but as McDonnell Douglas has continually stressed to us, safety is *always* the paramount factor. There is no doubt that if McDonnell Douglas had listened to the Dutch, and if the DC-10 floor had been strengthened to tolerate a hole in the fuselage of twenty square feet (as all wide-bodied floors must be able to do under the FAA's regulation introduced in July 1975), then the Paris crash would not have happened. The area of a DC-10 rear cargo door is a little over 14.5 square feet. If the Turkish Airlines' DC-10 floor had been reinforced and fitted with additional vents, the pressurized air in the passenger cabin would have escaped, leaving the floor intact and the control cables undamaged. At most, the passengers and crew would have suffered mild discomfort. However, McDonnell Douglas's decision to disregard the Dutch warnings—and later the warnings of the NTSB—relied on the fact that the DC-10 floor complied with the requirement of the FAA's current airworthiness regulations.

Most of the blame for the failure to do something about the extreme vulnerability of jumbo-jet floors must therefore lie with the FAA. Even after Windsor—when there would still have been time to prevent the Paris crash—the FAA was dilatory. In July 1975, by which time the danger could no longer be denied, the agency did finally order the modification of all DC-10, TriStar, and 747 floors.[1] But the delay undeniably cost the lives of 346 people.

NOTE

1. The floors have been modified to withstand the consequences of a hole up to twenty square feet appearing suddenly in the fuselage. In theory, the FAA's order affects only U.S. operators but, in practice, all airlines flying jumbo jets are expected to comply. Most DC-10s, TriStars, and 747s built since the beginning of 1976 have been fitted with strengthened floors: older versions are being modified as they are grounded for routine maintenance overhauls.

8

Compliance with
Service Bulletin SB 52-37

After the Windsor incident, McDonnell Douglas agreed to fix the cargo door problem by issuing service bulletins to owners of the DC-10. The following memorandum shows how long it took for each of the DC-10s to be modified in accordance with SB 52-37, which called for the addition of a support plate for the lock mechanism (see fig. 3.3, page 74).

HOUSE OF REPRESENTATIVES
SPECIAL SUBCOMMITTEE ON INVESTIGATIONS
COMMITTEE ON INTERSTATE AND FOREIGN COMMERCE

Washington, D.C., March 25, 1974

MEMORANDUM

To: Daniel J. Manelli, Chief Counsel
From: Mark J. Raabe, Staff Attorney
Subject: Failure to Achieve Prompt Compliance with McDonnell Douglas
Service Bulletin No. 52-37

At the time Service Bulletin No. 52-37 was issued, July 3, 1972, there were 39 DC-10's in service, all belonging to domestic carriers. Telephonic inquiries made to officials of these carriers on this date determined that in most instances long periods of time elapsed and millions of passenger miles were flown before the modifications called for in S.B. 52-37 were completed. As you know, as long as these modifications were not completed, the planes were in the generally same condition insofar as the aft cargo door was concerned as the DC-10 involved in the Paris accident.

Only five of the 39 aircraft were modified within 90 days of the issuance of the bulletin. Eighteen were not modified until 1973, with some of these not completed until June, eleven months after issuance of the bulletin. One plane was not modified until March 6, 1974, after the Paris crash.

Following are the results of inquiry showing each of the 39 planes and the date on which modifications under S.B. 52-37 were completed:

Manufacturer's Fuselage No.	Date Completed
American Airlines	
7	November 14, 1972
12	January 11, 1973
21	January 23, 1973
9	February 17, 1973
22	March 5, 1973
5	March 15, 1973
23	March 18, 1973
20	April 12, 1973
31	May 2, 1973
3	May 11, 1973
24	May 23, 1973
37	May 24, 1973
13	June 2, 1973
30	June 18, 1973

Manufacturer's Fuselage No.	Date Completed
Continental Airlines	
34	February 2, 1973
41	March 28, 1973
40	May 7, 1973
43	May 14, 1973
44	May 29, 1973
National Airlines	
18	November 2, 1972
19	November 9, 1972
38	December 2, 1972
16	December 6, 1972
14	March 6, 1974
United Airlines	
35	August 1, 1972
45	August 12, 1972
42	August 14, 1972
15	September 18, 1972
11	October 2, 1972
32	October 6, 1972
17	October 9, 1972
8	October 11, 1972
10	October 15, 1972
27	October 16, 1972
6	October 17, 1972
39	October 20, 1972
4	October 24, 1972
25	October 39, 1972
26	November 9, 1972

9

Conclusions of the U.S. Senate
Oversight Hearings and Investigation
of the DC-10 Aircraft*

FINDINGS, CONCLUSIONS AND RECOMMENDATIONS

The committee finds that the FAA, in dealing with the problem of the aft cargo door on the DC-10 following the June 12, 1972 incident, failed to make the proper regulatory action in seeking to ensure that a repeat of the near catastrophic door failure did not occur.

While the FAA's Western Region, the Flight Standards Service and the deputy administrator of the Agency had all agreed that the seriousness of the door problem warranted issuance of an airworthiness directive on the DC-10, the administrator, pursuant to an oral agreement with the president of Douglas Aircraft, decided to

---

* Reprinted, with editorial changes, from *Report on the Oversight Hearings and Investigation of the DC-10 Aircraft*, Committee on Commerce, Senate, June 1974.

address the problem with a company service bulletin, followed up by a telegraphic communication from the FAA to the airlines operating the DC-10. Testimony from the committee's hearings indicates that there was less than total compliance with all Douglas service bulletins dealing with the rear cargo door and that in at least one instance a U.S. registered aircraft was not modified pursuant to service bulletin 52-37 until after the Paris accident.

The committee however believes that the administrator was acting honestly and in good faith in dealing with the DC-10 problem and truly believed that the serious danger posed by the problem could be alleviated through the service bulletin process.

Nonetheless, we believe that when a serious or potentially catastrophic problem develops in a transport category aircraft, the public interest and safety requires strong regulatory action on the part of the Federal Aviation Administration. Issuance of airworthiness directives is the only proper way to proceed for several reasons. First, the AD has the force and effect of law and requires modifications or fixes be made by certain time deadlines if the operator wishes to continue to operate the aircraft. On the other hand, compliance with service bulletins is voluntary on the part of the operator and may be ignored if the operator chooses. Second, the AD is transmitted to all foreign governments by the United States notifying those governments whose airlines operate U.S. manufactured aircraft that an unsafe condition exists. Service bulletins, issued by a manufacturer, do not go to government officials and can result in governments being uniformed regarding potential safety problems. Third, the AD is a method by which the public becomes aware of safety problems in aircraft and focuses attention on the methods by which the government, the manufacturers, and the operators seek to alleviate the problem. In effect, the AD puts the government's action on the public record and can have a reassuring impact on the public which is anxious to see that the government is doing all possible to make air transport as safe as it can be.

On the other hand, the service bulletin process takes place quietly without public knowledge and in the DC-10 instance, communications between the FAA and the operators were given "for official use only" status, thus seeking to ensure that word of the safety problem did not become a matter of general public knowledge. We believe that subconsciously or otherwise, the manufacturer and the FAA sought, in dealing with this problem, to keep the bad news from the public in hopes of not blemishing the reputation of the DC-10 aircraft. While we share the concern that the DC-10 enjoy a high reputation for safety and reliability, we believe it is more important to deal with safety defects in an open and above board manner. If dealt with that way, then public confidence will be bolstered in the regulatory process and in the FAA which the public is entitled to believe exists to insure the highest safety standards in air transportation. Actions taken behind closed doors or actions which are kept off the public record only serve to create an impression of the regulator being unduly influenced by those he regulates. The regulatory process, particularly where public safety is concerned, should be totally above any suspicion that the government is compromising safety standards to accommodate industry. Confidence in government in general will not be bolstered by governmental actions which give the appearance of succumbing to private interests.

Fourth, and very importantly, the AD connotes a sense of urgency regarding a safety problem which might not attach to the issuance of service bulletins. The AD puts government inspectors on notice that a serious safety problem exists and that diligent oversight is required to make sure that the appropriate fixes and modifications are completed by manufacturer or operator. At the same time, the operators of aircraft, the airlines, are put on notice by an AD that quick and effective action is required and that delay or procrastination will not be tolerated. In effect, the AD is a red flag indicating that all is not well and that the government is requiring strong, affirmative, and public action to deal with the problem.

We note with approval a recent statement to the committee by the FAA Administrator following our hearings that the agency is changing the methods by which it deals with these problems.

> As part of the regulatory policy review, I have directed offices concerned to utilize airworthiness directive procedures in all future situations when a design change is needed to correct an unsafe condition. An agency order will formalize this directive and will emphasize that the use of general notices or other nonregulatory means should be discontinued for such a purpose.

We are also concerned by the FAA's failure to adequately follow up what action it did take in June of 1972 after the American Airlines incident. Following the agreement to proceed with service bulletins rather than airworthiness directives, the FAA, on its own initiative, failed to take action to ascertain whether the airlines were in fact providing the modifications called for in the Douglas service bulletins. We believe that under the circumstances and because of the urgency of the problem, that at the very least, FAA flight standards inspectors across the nation should have been checking to see if the airlines were properly incorporating the modifications pursuant to the service bulletins. This did not occur. In fact, the only followup was done by Douglas, which in August of 1972, informed the agency of the status of the modification program. Further, failure to follow up is noted in the July 7 letter from the administrator to the NTSB in which the administrator fails to inform the board of the issuance of Douglas service bulletin 52-37, the one the testimony indicates is the most important. Mr. Shaffer's letter adds strong weight to the proposition that he and/or the agency did not have adequate and thorough knowledge of the continuing steps being taken by Douglas to provide a fix that was "human" error proof.

Based on the testimony presented to us, the committee believes that the Douglas Aircraft Company did not take the kind of remedial action on the cargo door situation which circumstances warranted. While the company was quick to come up with the first fix and to notify the airlines via an "alert" notice of the fix, later and more comprehensive fixes were dealt with in a routine manner.

For example, when the modification contained in service bulletin 52-37 was arrived at on July 3, 1972, Douglas notified the airlines through a routine rather than an "alert" service bulletin. Yet, the modification called for in 52-37 was exactly what the safety board had earlier recommended in the way of a complete fix to the door problem. We do not understand why Douglas did not treat this service bulletin more urgently and can't help but wonder if whether this routine treatment was partially responsible for the failure of Douglas personnel to completely modify the THY and Lakers DC-10s which were then in the final stages of production prior to delivery.

Nonetheless, we feel compelled to point out, as did Douglas during the hearings, that proper door closure on the aft cargo door eliminates the possibility that the door would inadvertently come open in flight.

If proper procedures are followed by the airline employees charged with the responsibility of closing the doors, then the locking mechanism may be properly seated and the latch pins properly secured in place making it impossible for inadvertent door opening. While installation of service bulletin 52-37 makes it humanly impossible to improperly close the door, the trained, competent and conscientious employee would hopefully never force the door closed with undue pressure, thus providing an appearance of proper closure when in fact the door was not completely closed. While we do not know the circumstances, human and otherwise, involving door closure prior to the fatal Paris crash, we do point out that proper and secure closure was possible without the modifications provided for in service bulletin 52-37.

While there still exists much difference of opinion regarding the fix recommended by the safety board dealing with venting between the cargo compartment and the passenger cabin, we do not understand why this recommendation was given such low priority within the company and why it took more than a year before the company and the FAA dealt with the recommendation in an exchange of letters.

Finally, we find fault with the National Transportation Safety Board for failure to adequately follow up its recommendations on the DC-10 door problem and for its closing of the file on safety recommendation A-72-97 even though the FAA had not taken action initiating the proposed recommendation. The situation clearly indicates a lack of basic follow through to keep after FAA until the agency had met the objectives of the safety board's recommendations.

We hope that in the future, the board will be more persevering in pressing its recommendations with the FAA, if only in hopes that such pressure will vitalize the FAA with a sense of urgency when safety problems arise and will provide a force upon FAA which will keep the agency from letting problems drift.

The committee will shortly consider S. 2401, a bill which would reorganize the NTSB and which contains a provision requiring some written FAA action either affirmative or negative on NTSB recommendations within 90 days of their issuance. We believe that this requirement will assist NTSB in following up its recommendations and will require that the FAA deal with them in a public manner.

In summation, while it is a matter of sheer speculation whether different action by the FAA, by Douglas and by the safety board could have prevented the March 3 accident near Paris, we believe that the regulatory process designed to protect the public from unsafe conditions which could lead to air transportation disasters was circumvented in the DC-10 cargo door problem instance. We expect a better pattern of performance in the future.

THE 1974 PARIS CRASH

10

French Government Report on the 1974 Paris Crash*

Secretariat of State for Transport
Commission of Inquiry

Accident to
Turkish Airlines DC-10 TC-JAV in the Ermonville Forest
on 3 March 1974

FINAL REPORT
Translation of the Report published by the
French Secretariat of State for Transport

February 1976

---

* Reprinted, with editorial changes, from *Final Report, Secretariat of State for Transport, Accident to Turkish Airlines DC-10 TC-JAV in the Ermonville Forest on 3 March 1974.*

SUMMARY OF ACCIDENT

After a stop at Orly and a delay to its schedule because of the last minute embarkation of numerous passengers, TC-JAV [the designation code of this particular airplane] took off for London at 1132 hours.

Shortly after 1140 hours, when the aircraft had reached 12,000 feet during climb, the Air Traffic Control recorded a transmission in the Turkish language, partly covered by heavy background noise and accompanied by the pressurization warning and then the overspeed warning; at the same time the aircraft radar return split in two and the secondary radar label disappeared. Some seventy seconds later, the DC-10, flying at high speed and with a slight angle of descent, struck the treetops and disintegrated in the forest.

SEQUENCE OF EVENTS

The aircraft took off at approximately 113030 hours. Three or four seconds before 114000 hours, the noise of decompression can be heard on the cockpit voice recording, the co-pilot said: "the fuselage has burst" and the pressurization aural warning sounded.

At 114013 hours the controller who was following the progress of Flight TK 981 heard a confused transmission, a heavy background noise mingled with words in the Turkish language and the pressurization warning and then the overspeed warning. On the primary radar the aircraft echo split in two; one part (which may correspond to the parts ejected from the aircraft) remained stationary and persisted for two or three minutes; the second part, the echo of the DC-10 itself, continued on a path which curved to the left from heading 350° to heading 280°.

The various recordings (air/ground communications, cockpit voice recorder, flight data recorder) show that about 77 seconds elapsed between the time of decompression and the impact with the ground.

The flight data recorder show that, in the seconds immediately after depressurization, the speed of No. 2 [tail] engine dropped sharply and the aircraft turned to the left (9°) and went into a nose-down attitude. This nose-down attitude increased rapidly (down to -20°) and the speed increased (360 knots) though Nos. 1 and 3 engines had been throttled back. The pitch attitude then decreased progressively to -4° and the speed became steady around 430 knots (800km/hr).

The aircraft cut through the forest from east to west and caused damage over a rectangular area of 700m by 100m. The aircraft literally disintegrated on the subsequent impact at very high speed in the forest. The circumstances of the impact (disintegration at very high speed in the trees) were such that there was practically no fire. Jet Al fuel was used and there were about 23,500 liters on board at the time of impact.

AFT CARGO COMPARTMENT

The cargo door was closed at about 1035 hours by M. Mahmoudi, who has stated that he proceeded as usual, without any particular difficulties, and that he did not look through the view port, a procedure which he had seen but which he never carried out himself and the purpose of which he did not know. [The warning placard next to the view port is written in English, which he cannot read.]

In the absence of the station engineer who was on a training course at Istanbul, another engineer had been taken on for flight TK 981. After the closure of the rear cargo door on the left-hand side by M. Mahmoudi, no one saw that engineer or any other crew member inspect the lock pins by looking through the view port provided for that purpose, and their nonengagement was therefore not detected. Moreover, once M. Mahmoudi's work was completed, inspection would have entailed the positioning of equipment beneath the view port in order to gain access.

The loss of the door caused an almost instantaneous drop in the pressurization established in the cargo compartment beneath the passenger cabin floor. This excess pressure, added to the normal stresses on the floor, caused damage such that parts of passenger seats were ejected from the aircraft together with six passengers probably occupying two triple seat units in line with and above the cargo door. This damage was therefore clearly more substantial than in the Windsor incident, in which the initial floor loading was lighter.

Because all the horizontal stabilizer and elevator control cables are routed beneath the floor of the DC-10 and because of the priority assigned in this aircraft to each of these mechanical controls, the state of airworthiness of TC-JAV after the loss of the cargo door and the disruption of the floor structure must have been such that the crew were left with no means of regaining sufficient control of the aircraft.

MEDICAL INFORMATION

a) In the case of bodies recovered at the main accident site in the forest of Ermonville, there was a high degree of fragmentation (nearly 20,000 fragments were listed) associated with the violence of the impact.
b) On the other hand, the six bodies [passengers ejected when the door opened] found near St. Pathus were complete, although presenting fractures and serious visceral lesions.

DOOR TESTS

1. Latches closed, locking handle closed

The four lock pins were engaged behind the restraining flanges and prevented the opening of the latches, but they were only partially engaged. The ends of the lock pins were 1.6mm short of the rear face of the flanges [fig. 10.1].

The official adjustment documentation—*Maintenance Manual, Revision 4, January 1973*—stipulates that the ends of the lock pins should protrude for 6.35mm beyond the rear face of the flange. In consequence, with this adjustment the lock tube in its locked position was 6.35 + 1.6 = 7.95 mm short of the correct locked position.

2. Latches open

When the latches were open, the movement of the handle towards the closed position was stopped when the lock pins came up against the front faces of the flanges. [See Chapter Three for a description of the lock and latch system.]

Tests carried out on the same door, with varying adjustments of the extreme positions of the lock pins, showed that the force which has to be applied to the handle in order to force its closure depends on the extreme position (locked) to which the lock tube is adjusted.

When this adjustment is made in accordance with the manufacturer's requirements, i.e., when the ends of the lock pins protrude for 6.35mm beyond the rear faces of the flanges, it is physically impossible to force the handle even in the absence of the support plate for the vent door shaft.

On the other hand, when this distance of 6.35mm is decreased, the force required for forced closure also decreases. During the tests carried out (with the lock tube adjustment 7.95mm short of the cor-

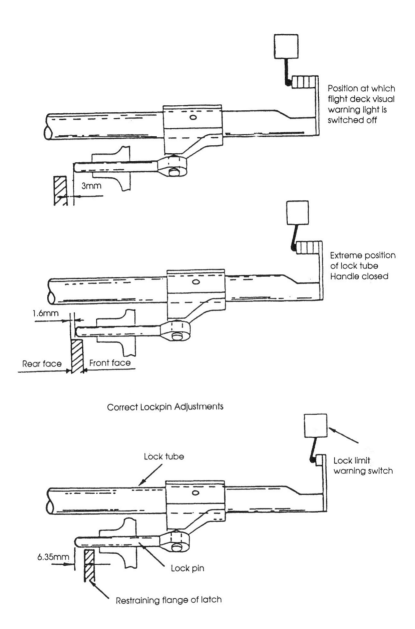

Position at which
flight deck visual
warning light is
switched off

3mm

Extreme position
of lock tube
Handle closed

1.6mm

Rear face   Front face

Correct Lockpin Adjustments

Lock tube

Lock limit
warning switch

6.35mm

Lock pin

Restraining flange of latch

FIG. 10.1
Lockpin Adjustments, Turkish Airlines DC-10

rect position) the handle could therefore be closed (and the vent door apparently closed) with a force of about 50 pounds.

This closure was possible only because of the deformation of the mechanism providing control transmission for the operation of the lock tube. The principal deformation affected the vent door shaft. The additional support plate specified by SB 52-37 [the McDonnell Douglas service bulletin issued after Windsor] was designed precisely to prevent such deformation. However, the additional support plate specified for the vent door shaft by SB 52-37 had not been installed. SB 52-37 does not list TC-JAV among the aircraft affected by the relevant modifications which ought to have been carried out by the manufacturer before delivery. As the result of an oversight by the manufacturer, the aircraft was delivered without the modification and the start of its application to the lock tube seems to show that work had begun on correcting the error retrospectively.

Because of the defective adjustment of the extreme positions of the lock tube, the handle could be closed without excessive force, although the latches were not completely closed. Defective closure could not be detected from the external appearance of the handle, vent door and cargo door, unless a visual inspection was made through the view port provided for that purpose.

When the lock tube was pushed towards the locked position, the [lock limit warning] switch switched off the flight deck warning light, although the ends of the lock pins were still 3mm from the restraining flanges. It should be noted, moreover, that according to the maintenance manual, the ends of the lock pins in the unlocked position must not be more than 2mm away from the flanges. In conclusion, the lock limit warning switch was defective and caused the flight deck warning light to go out when the latches were not necessarily closed.

COMPARISON WITH THE WINDSOR INCIDENT

Although the course of events and some of the causes are not exactly the same, nevertheless the Windsor incident presents points in common with the accident to TC-JAV:

- The latches were not fully closed and the latch lock pins were not in place.
- The flight deck warning light had gone out before effective locking had occurred.

- The altitude reached by the American Airlines DC-10 was of the same order as that of TC-JAV when the door opened and the two bolts (connecting the fixed part of the latch actuator to the door structure) failed under the same conditions.
- In the absence of pressure relief vents of adequate size between the passenger cabin and the cargo compartment, the sudden decompression in the cargo compartment caused damage to the cabin floor and its structure. This damage [Windsor] was less severe than in the case of TC-JAV in which the floor was more heavily loaded, but the functioning of the control cables was impaired in various ways, although it did not become completely impossible to control the aircraft.

Two recommendations had been issued by the [NTSB] investigators:

- modification to the locking system to make it physically impossible to position the external locking handle and vent door to their normal door-locked positions unless the lock pins are fully engaged;
- the installation of pressure relief vents between the cabin and the aft cargo compartment to minimize the pressure loading on the cabin flooring in the event of sudden depressurization of the cargo compartment.

The first of these recommendations had given rise to the modifications specified in Alert Service Bulletin 52-35 and in Service Bulletins 52-27 and 52-37 (in the case of 52-37, only a start had been made with its application to TC-JAV).

Other modifications of the door-closing system and methods of mitigating the effects of sudden depressurization of the aft cargo compartment were still under study at the time of the accident to TC-JAV.

Although there was apparent redundancy of the flight control systems, the fact that the pressure relief vents between the cargo compartment and the passenger cabin were inadequate and that all the flight control cables were routed beneath the floor placed the aircraft in grave danger in the case of any sudden depressurization causing substantial damage to that part of the structure. All these risks had already become evident, nineteen months earlier, at the time of the Windsor accident, but no efficacious corrective action followed.

11

Engineers Who Kill: Professional Ethics and the Paramountcy of Public Safety*

Thou shalt not kill.
—Exodus, Chapter 20

Engineers shall hold paramount the safety,
health and welfare of the public in the
performance of their professional duties.
—Engineers' Council for Professional
Development, *Code of Ethics*

The codes of ethics of a number of engineering professional societies[1] begin with language that states that engineers are required in their professional work to hold paramount the safety of the public. It is not difficult to appreciate why those in engineering should feel obligated to endorse such a statement nor is it hard to understand why it is generally placed first in the codes. For whenever we drive a car, or fly in an airplane, or take an elevator, or use a toaster, or cross a bridge, or do any one of a thousand daily tasks, we are relying upon the work that engineers have done, entrusting our lives to the products of their special skills. As a profession—or perhaps, more properly, as a profession in the making—engineers want to alert themselves to the responsibilities created by the public's reliance upon them. A world in which the safety of the public was routinely

---

* Reprinted with editorial changes by permission of the author. Copyright ©
1981 by Kenneth Kipnis.

subordinated to other concerns would likely be a very dreadful world indeed. And so members of the community need to be assured that engineers—who can both cause and avert grave public damage—have the same concern for safety that others are expected to have, especially those others in positions of trust. Andrew G. Oldenquist and Edward E. Slowter have seen the codes as emerging from the stewardship bestowed upon the profession by society:

> The public's health, welfare and safety depend on the knowl-
> edge, competence, and integrity of all engineers responsibly
> involved in any given engineering activity. These engineers do
> not merely possess special knowledge and capabilities; like
> judges and physicians they have exclusive stewardship over
> their special knowledge. Our society has said that one must be
> an engineer in order to do certain activities, and that only other
> engineers are in a position to evaluate the quality of the engi-
> neer's performance. Professional integrity is especially vital for
> any professional group that is self-regulating on expert matters
> and to whom society has given stewardship over important
> knowledge and activities. A code of ethics serves to remind
> individuals how important integrity is in a self-regulating pro-
> fession. It lays out the specific matters deemed most impor-
> tant in the collective wisdom of the profession, and in solemnly
> promulgating and enforcing the code, notice is served that the
> elements of the code are to be taken seriously.[2]

Professionalism thus involves more than simply expertise; it involves a public commitment to some set of significant social values. This commitment forms a part of the reason for public reliance upon the profession as the means by which certain skills and knowledge are applied within the community. Hence, if engineers wish to be recognized as professionals, it is necessary for them to declare publicly their allegiance to certain norms that define their commitment to the public.[3]

These norms that are articulated in codes of ethics are of two basic types: what we may call *ideals of the profession* in the one case and *principles of professional conduct* in the other.[4] The ideals are aspirational in nature, calling attention to the central goals of the profession. They are important because they provide guidance in the development of an exemplary professionalism. But while they can be flouted and ignored, they cannot, strictly speaking, be violated. You can choose not to follow the path indicated by a statement of the profession's ideals, and if you do you may never become an exemplary professional. But that does not mean that you have broken some rule. Principles of professional conduct, however, do provide criteria for judging when a professional has fallen culpably short of acceptable standards for the profession. Unlike an ideal, a principle of professional conduct can be violated. It specifies conditions under which a practicing professional violates the public trust undertaken when he or she became a professional. While the failure to realize one's full potential as a professional may be lamentable, the violation of a principle setting forth the minimum requirements for acceptable professional practice is a very serious matter, one that should merit the attention of all professionals with a stake in the quality of their profession.

When an engineering professional society asserts that engineers shall hold paramount the safety, health and welfare of the public, it is not at first glance clear whether what is being said represents an ideal to which engineers ought to aspire or whether it represents a principle that engineers violate at the cost of their professional integrity. But not much reflection is required to see that both ideals and principles are involved. It is arguable that the very highest achievements in engineering have been those in which the ambient levels of safety, health and welfare have been improved. And it may well be that engineers share a common and firm commitment to continued progress in these areas. In these two ways, the profession may have as an ideal a devotion to the furthering of these goods. And yet, because it is an ideal, the engineer who cannot point to any way in which he or she has contributed to improving safety, health or welfare is not thereby convicted of unprofessional or unethical conduct. If a charge of unethical conduct is to be warranted, there must be a principle of professional conduct that can be violated. If engineers are required to hold paramount the safety, health and welfare of the public, does that mean that there are specific acts that engineers are not permitted to perform? Is there a point at which the conduct of an engineer is professionally unacceptable, given the language of the codes? The answer seems to be that, though engineers are not culpable if they merely fail to make contributions to public safety and the rest, they are culpable if they act as a clear menace. To hold paramount the safety, health and welfare of the public precludes, at least in most cases, acting so as to undermine those values.

In what follows, I shall begin the task of fleshing out what is meant by "acting as a menace." The aim will be to articulate *some* of the principles of professional conduct that are implicit in the provision that engineers hold paramount the safety of the public. I do not propose to cover all of the principles but, rather, I shall confine myself to those addressed to the very clearest and most pressing cases of professional misconduct in engineering. Accordingly, because the concepts of health and disease are both controversial and complex, I shall not discuss them at all in what follows. And with respect to the remaining notion, safety, I shall consider only those incursions upon safety that are reflected in an increased expected incidence of death or a higher mortality rate. For a concern with public safety has to mean, at a minimum, that one will not participate in the killing of others. To be as clear as possible, I am fully aware that there are more ways of being an unethical engineer than are discussed in what follows. The claim I make is that the principles set out below, as explained, will pick out only serious cases of professional misconduct. This work will show, I trust, that the derivation of the remaining principles is a task worth undertaking. We are concerned here, at the outset, with those engineers who represent the most serious threat to public safety: those who kill.

Leaving engineering, let us consider a clear case of someone acting as a menace: a man who knowingly fires a rifle at the crowded bleachers at a baseball game. The sniper subjects a thousand innocent unsuspecting people to a risk of death and, even if no one is killed, the public is endangered in a most serious way. Even if the rifleman has some innocent purpose—say he is testing the rifle or trying to kill a mosquito—he is nonetheless a palpable menace to public safety. Reasonable people do not want others to act in this way. So it would seem that a principle committing the engineering profession to the paramountcy of public safety would prevent one from imposing the risk of death upon others. We might thus consider P^1 as an appropriate principle of professional conduct for engineers:

P^1: Engineers shall not participate in projects that subject others to a risk of death.

But this principle will not do. For there are situations in which it is morally acceptable to impose a risk of death upon another. A surgeon, for example, may subject a patient to a 25 percent risk of death in an operation. If the operation provides a 75 percent chance of curing a condition that is 100 percent fatal, then, even though the surgeon kills the patient on the operating table, she is not thereby culpable. One must consider the effects of the risky operation upon the patient's "background" risk. The patient is safer with the operation than without it. Similarly, a community may be plagued by mosquitoes carrying a disease that kills ten persons per year. The only way to eliminate the disease is to spray a chemical into the air that will kill the mosquitoes. But the spray can be expected to kill one of the thousand members of the community.[5] Since, in this case, the community would enjoy an improved level of safety despite the single expected fatality cause by the spray, the killing would, it seems, be justifiable. And so it would seem that a commitment to the paramountcy of public safety would imply a commitment to P[2]:

P[2]: Engineers shall not participate in projects that degrade ambient levels of public safety.

But this won't do either. The patient of the surgeon may not have a fatal disease but, rather, one that causes great discomfort. If the patient is willing to assume a 10 percent risk of death (as a consequence of the operation) in order to have a 90 percent chance of relief from his painful but non fatal condition, then though the patient is killed by the operation, the surgeon is not thereby culpable. This means, of course, that the patient has to have available to him information about the risks and benefits associated with the surgeon's treatment. Thus, though the surgery increases rather than reduces the patient's overall risk of death, the surgeon's action may be morally acceptable if full and accurate information has been made available to the patient and the patient consents.[6]

However, in the chemical spray case, securing consent can be more complicated. Let us consider three variations.[7]

Case 1: In this case and the ones to follow, the mosquitoes are merely very annoying. They do not carry a fatal disease. The spray will harm none of the thousand members of the community except for one: this is you. Because of a congenital sensitivity you will die. Leaving the community is not a possibility and you know of your special vulnerability as does everyone else.

It would seem that, in a case such as this, you ought to have a veto over the spraying if it means you will lose your life if it takes place. The others will enjoy the benefits: you will assume the gravest costs. If someone were willing to lay down his life so that others might enjoy a mosquito-free existence, then *perhaps* the spraying should occur. But no one should be required to die merely to improve somewhat the lot of others.[8] Here the consent that is crucial is the consent of the individual placed at risk. Thus, information about the nature of this increased risk, if it is available at all, should be available to that person.

Case 2: The spray will have no effect on 997 of the thousand members of the community. For three members there is, for each, a one-in-three chance that he or she will die. (There is thus a 1/27 chance that all three will die and a 8/27 chance that none will.) You are one of the three.

Once again it would seem that those placed at heightened risk ought to have a veto. But should the veto be exercised individually or collectively? Suppose, to focus matters, the other two are willing to consent to the spraying but you are not. Should you be compelled to go along with the majority in this case? I would think not. That two others are quite willing to risk their lives seems to have nothing whatever to do with your right to risk or not to risk your life as you see fit. And so in this case it would appear that unanimous consent of those placed at increased risk would be required if the spraying were to be undertaken in a morally acceptable way.[9]

Case 3: Each of the thousand has a 1/1000 chance of dying from the chemical spray.

By analogy with case 2, each of the thousand is placed at heightened risk and should therefore have a veto over the spraying. That means that each of the thousand must be informed (or waive the right to be) and must consent if the spraying is to take place in a morally acceptable way. And so it would seem that a commitment to the paramountcy of public safety would imply a commitment to p^3:

P^3: Engineers shall not participate in projects that degrade ambient levels of public safety unless (1) appropriate information concerning those degradations is made generally available and (2) each of those placed at increased risk consents.

P^3, though it may be sound in some ultimate sense, appears to pose serious theoretical and practical problems. These center around the condition that consent be secured from each of those placed at increased risk. In the first place, not everyone at risk would be in a position to grant or withhold consent. It is possible for engineering projects—nuclear devices, for example—to jeopardize unborn generations, persons who cannot consent because they do not yet exist. Still other people may be unable to consent because of incompetence: children, the retarded, the mentally ill, the comatose. Consent from others may be invalid because of duress, undue influence, or the high cost of withholding consent: consider the plight of draftees ordered to use an unreasonably hazardous piece of military equipment. More generally, it should be remembered that consent is frequently exercised in some sort of political process. But clearly some political systems do not provide everyone involved with a real capacity to withhold consent. The problem plainly arises within tyrannical forms of organization but it may exist in other contexts as well. I am uncertain, for example, that each of us has consented to the "acceptable" risk levels allowed by the Atomic Energy Commission. These problems, concerning the conditions under which valid consent may be said to have been given, have been deeply perplexing to specialists in legal and political theory.[10] Protests over nuclear power installations suggest that the problems are unsettled in the political arena as well. Thus, given the present uncertainty surrounding the issues here, it would be unreasonable to expect engineering professional associations to make responsible authoritative judgments on issues involving the adequacy of consent. And even if we were sufficiently clear about the issues, it would not be plain that engineers would have responsibility *as engineers* to ensure that risks not be imposed without proper consent: if engineers have this responsibility, they may have it as citizens, sharing it with others. What we can say is that, regardless of consent, information about risks should not be withheld from the public. Once the information is made generally available, the question of consent can be raised. The question cannot even be taken up if the risk is concealed.

And so we are brought to our First Principle:

> *First Principle:* Engineers shall not participate in projects that degrade ambient levels of public safety unless information concerning those degradations is made generally available.

Notice that this principle *does not require* that engineers convey information about hazards to either the public or appropriate public officials. (Engineers could, however, comply with the principle by doing so.) The principle requires merely that engineers not participate in the covert degradation of public safety. The obligation not to participate in the killing of others is much stronger than the obligation to prevent others from killing. But the line between these two obligations may be blurred in cases where one discovers that one *has* participated in the covert degradation of public safety. It is now too late to withdraw: one's job is completed. In these cases, where one shares responsibility for the creation of a life-threatening hazard, the

principle does create a duty to warn those at risk. This is not to say
that engineers are never obligated to publicize hazards that they
did not help to create. To specify the conditions under which engineers have such obligations would require analysis and principles
that are beyond the scope of this paper.

If a professional association were to adopt the First Principle as
a principle of professional conduct, one would expect it to be applied
in two distinct ways. In the first place, the principle could be used to
warrant a decision to censure or discipline the engineer who violates it. What would need to be shown in these cases would be that
the engineer either *understood or should have understood* that the
project he or she was working on had increased the risk of death
from some above pre-existing background levels, and that information about that increased risk had not been made generally available.
In the second place, the principle could be used to warrant a decision
to protect or defend an engineer who has been required by his or
her employer to violate the obligation specified by the First Principle
but had refused to do so.[11] What would need to be shown in these
cases would be that the engineer *had made a reasonable professional
judgment* that the project that he or she was working on was likely to
increase the risk of death from some above pre-existing background
levels, and that information about that increased risk was not to be
made generally available. Professional associations must be careful
not to impose a superhuman standard of omniscience before deciding
to come to the assistance of the engineer endeavoring to comply with
the First Principle. Engineers cannot be required to know for certain
that public safety will be degraded. All that can be required is that
the engineer has made precisely the same type of informed and professionally responsible judgment that he or she has been trained to
make and hired to make. We rely upon that developed capacity for
judgment all of the time. One must be suspicious of efforts to call
that judgment into doubt and to demand more only when the safety
of the public is being judged to be at heightened risk. As the likelihood of a heightened risk increases, and as the magnitude and probability of the damage increases, the needs for support and involvement of the profession become more pressing.

For clarity, some commentary on the language of the First Principle is in order.

"Engineers." This would include all engineers: not just those
belonging to professional associations, state registered and otherwise. It is engineers in general who are capable of eroding public
trust in engineering; not simply engineers who are members of soci-

eties. Possibly there would be differences in the way the principles would be enforced (non-members could not be expelled from associations), but there might not be differences if action by professional associations took the form of public censure of unethical engineers or employers who require engineers to violate the principles of professional conduct. This latter type of censure could at least serve to warn engineers against working for such employers. These warnings are particularly appropriate in the case of engineering students, interviewing for their first job. Engineering professional schools might refrain from assisting these censured employers in arranging interviews with graduating students.

"[S]hall not participate in projects." There is room for some flexibility in interpreting this clause. Construed broadly, it might mean that the practicing engineer could no longer work for the employer once he or she had formed the judgment that the organization was degrading ambient safety levels without making the facts known. Construed narrowly, it might mean that the engineer is required only to withdraw from a particular subtask of a larger activity if that is where the hazard is localized. I would expect that discretion here could best be left to the practicing engineer, at least until the difficult problems surrounding the concept of "complicity" were worked out. There is also the question of when withdrawal is required. The engineer may understand merely that there is some likelihood that a risk of unknown probability or magnitude may be imposed by a project just in the planning stage. Or it might be understood that the organization is actively involved in the degradation of public safety. It should be remembered that as projects get underway, it becomes more costly and difficult to alter course. For purposes of *protecting* employed engineers, *withdrawal would be justified* at the point where it becomes clear that the employing organization is likely to impose covert risk. For purposes of *disciplining* engineers, *failure to withdraw would be culpable* at the point where covert risk is imposed.

"[I]nformation." This would include both information about the type of threat imposed upon the safety of the public and the magnitude and probability of the risk.

"[G]enerally available." With few exceptions, this condition would be satisfied if the information were conveyed to a governmental authority with responsibility for the type of threat imposed by the project: the FAA with respect to aviation safety, the EPA with respect to environmental hazards, and so on. The most glaring cases seem to be those in which organizations endeavor to keep such

information concealed from people with an interest in it. Such an effort typically requires not merely that one refrain from making it available to appropriate governmental officials but that one also prevent it from leaking out. Certainly if an organization fires or threatens to fire an engineer for making available appropriate information about a threat to the public safety, that would be a conclusive reason to believe that the organization is concealing the facts itself. Complications arise where government officials ignore their responsibilities, abuse their authority, or are major perpetrators in the concealed degradation of public safety. Suffice it to say that the test is the availability of the information to those placed at increased risk. Where it is not reasonable to believe that the appropriate information is available, this condition is not met.

One major exception to the First Principle involves the intended harmful effects of national defense weaponry. It is persuasively arguable that mortal enemies may not be entitled to a full account of the risks to which they might be subject. Some degree of secrecy may be permitted here. For this reason, the First Principle does not require that organizations producing weapons for national defense purposes make generally available information about the risks these weapons may impose upon enemies. Other risks however—those imposed upon ourselves or upon those who handle the weaponry—must not be concealed.

Let us look at a specific case to get a feel for the way this principle would apply. By June of 1972 it had become quite clear to F. D. Applegate, director of product engineering for Convair, that the McDonnell Douglas DC-10 upon which he was working (Convair was a major subcontractor) was a seriously flawed aircraft. The immediate problem was a badly designed latching mechanism for the cargo doors. Though corrective "band-aids" had been applied, the mechanism was far from "fool-proof" and Applegate anticipated that the cargo doors would continue to open when the plane became airborne. There had already been two such failures. But the failure of the cargo door was not itself the central cause of concern. As the cargo compartment lost pressure, the floor of the pressurized passenger compartment above would collapse downward into it. Because the main control lines to the rear of the DC-10 were positioned just below the cabin floor, such a collapse could be expected to disable controls to the rear of the plane. One of the engines is in the rear as are many of the control surfaces. In a memorandum expressing his concern to the manager of Convair's DC-10 Support Program, Applegate wrote that the plane had "demonstrated an inherent susceptibility to catastrophic failure." "It seems to me inevitable that, in the twenty years ahead of us, DC-10 cargo doors will come open and I would expect this to usually result in the loss of the plane." Applegate concluded his memorandum with the suggestion that certain changes be made in the design of the cabin floor that would prevent catastrophic failure. In his reply to Applegate's memo, the program manager did not take issue with any of Applegate's factual claims but made it clear that no effort would be made to correct the problem or even to report it to McDonnell Douglas. Needless to say, the substance of Applegate's memo was not made available to the FAA.[12]

Upon receiving his reply from the program manager, it should have been clear to Applegate that the continuing introduction of DC-10s into the stream of commerce represented a substantial degradation of the ambient safety levels for air travelers. One must assume here, as seems reasonable, that Applegate understood that such a degradation would occur whenever an aircraft with an "inherent susceptibility to catastrophic failure" was introduced. It was likely clear as well to Applegate that information about the nature and magnitude of these risks was not being made available to the FAA. Indeed, Convair was forbidden, under its contract with McDonnell Douglas, from contacting the FAA about the matter.[13] To put the matter bluntly, Applegate was almost certainly aware that he was involved in a project that would inevitably kill people, people who were not aware, who could not be aware, of the risks to which they were subject. It appears that Applegate did not quit the DC-10 Program.[14] If all this is correct, Applegate understood that he was participating in a project that degraded ambient levels of public safety without information about that degradation being generally available. Though there are others who would likely share blame in this case—engineers, managers, government officials—Applegate probably deserves some of it. Twenty-one months after Applegate sent his memo, a Turkish Airlines DC-10 lost a rear cargo door 11,500 feet above Coulommiers, near Paris. As Applegate foresaw, the cabin floor collapsed, the controls to the rear of the plane were disabled, and the aircraft, with 346 persons on board, hit the ground at 497 miles per hour. It was the first crash of a loaded jumbo jet. To say there were no survivors would be to understate the dimensions of this tragedy.

There is an important type of case, however, that the First Principle will not reach. In 1972 a train on the Bay Area Rapid Transit system (BART) overran the Fremont station and crashed into a sandpile.[15] There were injuries though no one was killed. An investigation showed that an oscillator had failed in the train's control system and that, as a consequence, the system had read a signal to slow down as a signal to speed up. Two facts make the case an interesting one from the present perspective. First, the introduction of the BART system probably represented a very substantial improvement in the ambient levels of safety in the area. Public transportation—even defective public transportation—is substantially safer than the automobile. So in this case it would have been persuasively arguable that there was no degradation of ambient safety levels; that, on the contrary, there was improvement despite the crash. Second, it is an accepted standard in the engineering of public transportation systems that they be designed so that when a part fails, as might be expected in the life of the system, that failure does not precipitate a life-threatening occurrence; that they be designed, in other words, to be "fail-safe." Knowing that oscillators fail, the engineers working on the control system should have designed it so that when the failure occurred, the train would slow down or stop—not accelerate and crash. In this case that standard was violated and the "Fremont Crash" was the result. It would appear that the engineers who violated that standard could be held to be responsible for the hazards that resulted. There might be engineering malpractice here. But the First Principle *cannot* be appealed to in condemning what the engineers did. This is because there was no degradation of ambient safety levels as there must be if the First Principle is to apply. One cannot justify a needless, negligent threat to public safety by pointing out that one has saved several other lives. And yet, in the Fremont Crash, precisely the latter type of fact foils the applicability of the First Principle. Accordingly, another principle must supplement the first.

> *Second Principle:* Engineers shall not violate accepted or acceptable standards of professional practice in cases where the safety of the public would thereby be placed at heightened risk.

Needless to say, the Second Principle creates a special obligation to apprise oneself of all applicable professional standards in projects that affect the public safety. It creates as well an obligation to withdraw from a task if it turns out not to be possible to adhere to accepted or acceptable standards in the course of it. Although there are many occasions in which it might well be appropriate to cut corners, the Second Principle provides a guarantee that engineers will not do so where the public safety is thereby placed at greater risk.

Again some commentary is indicated.

"[A]ccepted or acceptable standards of professional practice." These are and would continue to be defined by the engineering profession. Although the profession may and does define standards in various areas, it should be open to engineers to depart from these if they can demonstrate that the ones they are adhering to are at least as respectful of public safety as the accepted ones. The burden of proof, of course, is upon the engineer who departs from the accepted standards.

"[C]ases where the safety of the public would thereby be placed at heightened risk." Here the risk created by the violation of the safety standard is to be compared with the risk that would exist in the absence of the violation. In the Fremont Crash, we would be comparing the risk imposed by the defective train with the risk that would be imposed by the train had it been properly designed. We are not comparing—as we would with the First Principle—the risk imposed by the system with the pre-existing background risks imposed in the absence of the system.

The upshot of this chapter so far has been to sketch why and how engineering professional associations might take an expanded role in setting the conditions of professional practice. It is time now to consider some of the main objections to this view.

"When engineers kill others, that is a legal problem: not a moral one. It is a problem for the legal system: not for the professional association. Engineers need take into account only what the law requires. There is no need for professional associations to take action here."

In practice, there are several reasons why the criminal law is not (and, for the foreseeable future, will not be) used against engineers who participate in the concealed degradation of public safety. Prosecutors are under substantial pressure to attend primarily to a very different type of wrongdoing: street crime. The inadequacy of their resources has meant that they ignore some crimes on the books and offer substantially reduced sentences to the wrongdoers who are caught in order to persuade them to plead guilty and waive their right to trial. Trials exact a frightful toll on the prosecutor's meager resources. If a prosecutor were to proceed against an engineer, it would be expected that the employing organization would come to the defendant's rescue. (Corporations worry that the employee will try to shift the blame to them.) Once the corporation has entered the legal arena, the prosecutor can count on very pressing demands being placed upon the scant resources of the office, which is simply not geared to handle this type of case. The Ford Motor Corporation spent many times what the prosecution did to defend itself in the Pinto "exploding gastank" case. Additionally, it is often difficult for prosecutors to be fair in including as defendants some of the participants in a fatal project but not others. For these reasons and others, engineers who participate in the covert degradation of public safety have little to fear from the criminal law. They are effectively insulated from its requirements.

"What about the civil law? Can't engineers be sued and held liable for damages? Can't they be made to pay for the deaths they help to bring about?" Damages in the type of case we have been considering are typically very substantial; far, far beyond the capacity of even very successful engineers. The awards in the DC-10 case have exceeded 100 million dollars already. Working engineers are what plaintiffs' attorneys call "judgment proof." It does not make sense to sue them because they cannot begin to pay for the harm they have done. And so, as a matter of fact, engineers are insulated from the obligations imposed by the civil law as well.[16]

Though engineers may not be held criminally responsible for the deaths of others, though they may not be held civilly responsible, they may yet be held morally responsible. Professional associations have both the competence and the standing to make authoritative judgments in these cases. It is not likely that the law can do an acceptable job in securing a tolerable level of ethical behavior among engineers. But the profession has, as part of its socially bestowed stewardship, the obligation to see to it that engineers show a civil respect for the public safety. And, acting through organizations, the profession can secure a substantial level of compliance with appropriate principles of professional conduct.

"But aren't the employers of engineers the ones with responsibility for these deaths? Doesn't that fact make a difference?" Although the corporation may own what the engineer produces, the engineer does not cease to have responsibility for the reasonably foreseeable consequences of his or her acts. Certainly if a corporation paid an engineer to test-fire a rifle at a crowd in a baseball stadium, the engineer would be responsible for the deaths. The fact that he was paid to do the job is simply irrelevant to the question of his responsibility (though it would be relevant to the employer's). Moral responsibility for one's actions is not given up when one becomes an employee. The fact that engineers are not criminally prosecuted and the fact that it has been corporations

that have faced civil liability have, I believe, tended to blind engineers, like Applegate, to their own moral responsibility for the deaths of others.

"But don't engineers have a competing duty of loyalty to their employers and to their co-workers?" Loyalty alone is not a virtue, for it can be blind and misplaced. The loyalty that one may find in a gang of cutthroats is a part of what makes them wicked. Loyalty becomes a virtue when it is coupled with discernment in the choice of its object. The reminder that engineers hold paramount the safety of the public can be a great help to young engineers in the development of that discernment.

"What about the working engineer with a family to support? One cannot expect these engineers to walk off their jobs." I do not think that obligations to family are strong enough to permit one to take a job in an organization that kills. They would not permit one to take a position as a highly paid assassin for the Mafia. If I have to kill others in order to provide the very best for my family, maybe my family should learn to do with second best.[17] Certainly I do not do my best for my children when I help to contribute to the world more death, more injury, and more sickness. In any case, an enhanced role by professional associations in the protection of ethical engineers will have the effect of reducing the personal costs of ethical behavior. These painful dilemmas may thus become less common than they have been.

"What about the problem of enforcement? Even if professional associations merely censure engineers and employers of engineers, they will open themselves up to libel suits that may bankrupt the associations." Here, engineering professional associations have much to learn from other organizations and professions that have grappled with similar problems. The American Association of University Professors regularly censures institutions of higher education for violations of the academic freedom of faculty. Consumers Union regularly excoriates manufacturers for the shoddy quality of their goods. Both have apparently solved the problem of libel suits.

But is the instituting of a mechanism for censuring engineers and their employers sufficient to secure adequate compliance with principles of professional conduct? This is a question that can best be answered by experimentation of the sort that engineers are supposed to excel at. The design of an effective mechanism for securing among professionals an appropriate level of respect for the public safety is very much like other problems in engineering. Engineers can do what needs to be done to give concrete expression to their commitment to the paramountcy of public safety. Those of us who look to engineering as a major source of human betterment would urge no less.

NOTES

1. In addition to code of the Engineers' Council for Professional Development, similar language also appears in the codes of the National Council of Engineering Examiners, the Institute of Electrical and Electronics Engineers, and the National Society of Professional Engineers. The statement is also part of the *Uniform Code of Ethics of the Engineering Profession*, draft, by Andrew G. Oldenquist and Edward E. Slowter, presented at the meeting of the American Institute of Chemical Engineers, August 17-20, 1980, in Portland, Oregon.

2. Andrew G. Oldenquist and Edward E. Slowter, "One Code of Ethics for All Engineers," presented at the meeting of the American Institute of Chemical Engineers, August 17-20, 1980, in Portland, Oregon.

3. The relationship between professionalism and the obligations imposed by the codes of ethics is more fully explained in Kenneth Kipnis, "Professional Responsibility and the Responsibility of Professions," in the proceedings of the 1978, University of Dayton Colloquium on Collective Responsibility in the Professions, *University of Dayton Review*, summer 1981, and in Joseph Ellin, Michael S. Pritchard, and Wade Robison, eds., *Ethics in Business and Professional Life* (Clifton: Humana Press, forthcoming).

4. The distinction is drawn similarly in codes of ethics from other professions. Consider, for example, the American Bar Association *Code of Professional Responsibility*, which distinguishes between Ethical Considerations (ideals) and Disciplinary Rules (principles). EC 2-25 provides, among other things, that every lawyer "should find time to participate in serving the disadvantaged." Lawyers who fail to find such time are not regarded as having fallen short of acceptable professional standards. DR 9-102 provides, however, that lawyers shall maintain separate bank accounts for the funds of clients. Lawyers are regularly disciplined for violation of that rule.

5. Problems would arise if the members of the community vastly preferred to die of the mosquito-borne disease than to be killed by the spray. We will pass over these. They would arise as well if the people at risk from the disease were not the same as those put at risk by the spray. (Say only adolescents were endangered by the spray while only the very aged were subject to the disease.)

6. Here, allowance must be made for the patient who does not want to know the risks but is willing to let the trusted surgeon act using his or her

best judgment. It is enough that the information simply be available; the patient need not actually be informed. Not permitted are the exaggeration of benefits and the failure to make available information concerning the risks. The concept of "informed consent" has been fairly well developed in the field of medical ethics. See, for example, the anthology *Contemporary Issues in Bioethics* by Tom L. Beauchamp and LeRoy Walters (Belmont: Wadsworth Publishing Company, 1978), pp. 143-168 and pp. 430-441. It has not however assumed a central place in engineering ethics although Robert Baum has appealed to informed consent in his "The Limits of Professional Responsibility" in *Ethical Problems in Engineering*, 2nd Edition, Volume One, Albert Flores, ed. (Troy: Center for the Study of the Human Dimensions of Science and Technology, 1980), pp. 48-53.

7. I owe these examples to Stephen H. Unger.

8. This is an important part of the position taken by John Rawls in his highly influential *A Theory of Justice* (Cambridge: Harvard University Press, 1971). The obligation to treat people as ends in themselves—not as means only—precludes imposing "upon them lower prospects of life for the sake of the higher expectations of others," p. 180.

9. Things might be different if everyone in the community had, with due deliberation, agreed beforehand to go along with some majority-rule procedure in cases such as these.

10. See, for example, the selections in Lyman Tower Sargent, ed., *Consent: Concept, Capacity, Conditions, and Constraints* (Wiesbaden: Franz Steiner Verlag GmbH, 1979).

11. Engineers in this position have historically been disciplined by their employers, fired, and even blacklisted for refusing to comply with orders requiring them to violate professional ethics. An important court case—*Pierce v. Ortho Pharmaceutical Corp.*, 417 A.2d: 505 (N.J. July 28, 1980)—has recently provided a legal basis for professional associations to assist employed professionals caught in such dilemmas.

12. The Applegate Memorandum and the reply to it are reprinted in Paul Eddy, Elaine Potter, and Bruce Page, *Destination Disaster* (New York: Ballantine Books, 1978), pp. 274-80. Chapter 6 contains the Applegate memorandum.

13. Ibid., p. 268.

14. Ibid., p. 426. Applegate is still a senior executive with Convair.

15. On the BART case, see Robert M. Anderson et al., *Divided Loyalties* (West Lafayette: Purdue University, 1980). For additional information I am indebted to Stephen H. Unger and Robert Bruder, both of whom were involved in the case.

16. These considerations have not, however, prevented corporations from ascribing responsibility to underlings when catastrophes occur. McDonnell Douglas blamed the Paris DC-10 crash on a baggage handler at Orly Airport. Eddy, Potter and Page, *Destination Disaster*, p. 209.

17. Perhaps things might be different if the engineer's family faced starvation. But in practice this is not an issue.

12

Whistleblowing, Ethical Obligation, and the DC-10

Whistleblowing is an important ethical issue and an instructive one since it often highlights a conflict between acting for the benefit of others or in one's personal interest. Whistleblowers attempt to disclose wrongdoing within or by an organization to the public, government, media, or members of their organization other than their immediate superiors. The purpose of the disclosure is to prevent the wrongdoing from continuing. Whistleblowers risk their personal well-being because organizations often retaliate against them.[1] Since whistleblowing has the potential to benefit society but often involves personal sacrifice, a crucial question is: When, if ever, are businesspeople or engineers ethically obligated to blow the whistle on their employers?

This chapter divides the treatment of the question into four parts. First, I present a set of conditions, proposed by Richard DeGeorge, which claim to establish when whistleblowing is ethically permissible and ethically obligatory. I argue that these conditions have a major fault: in almost all complex cases an employee would not be obligated to blow the whistle. This flaw keeps the conditions from making adequate moral demands on employees and overprotects companies. The second part of the chapter employs DeGeorge's conditions to discuss whether Dan Applegate, a high-ranking engineer employed by the Convair Division of General Dynamics, was ethically obligated to blow the whistle in the DC-10 case of 1972 to 1974.[2] This application of the conditions to a case illustrates the problem with them.

Part 3 presents my approach to ethical obligation and whistleblowing. I argue that engineers and businesspeople have an ethical obligation to blow the whistle more frequently than DeGeorge's conditions indicate. This view provides more protection for users of products and services. The final section of the chapter applies my position to the DC-10 case.

PART I

How are we to determine whether businesspeople and engineers are ethically obligated to blow the whistle on their employers? One response to this question has been to try to set out conditions or criteria for whistleblowing. Richard DeGeorge claims there are five conditions which determine the moral status of whistleblowing. If the first three are satisfied, whistleblowing is permissible; while if all five are satisfied, whistleblowing is morally obligatory.

The first condition is: "The firm, through its product or policy, will do serious and considerable harm to the public, whether in the person of the user of its product, an innocent bystander, or the general public."[3] By "serious and considerable harm," DeGeorge means serious harm or danger to life or health. He wants to limit whistleblowing to situations that threaten death in order to create a set of clear-cut cases to act as a paradigm. He also believes that such a limitation would avoid unnecessary harm to companies and would keep whistleblowing a rare occurrence, which would make it more effective.

The second condition states that, "Once employees identify a serious threat to the user of a product or to the general public, they should report it to their immediate superior and make their moral concern known." DeGeorge thinks that people have a moral obligation to prevent harm to others if they can do so at relatively little cost to themselves. Presumably, reporting a threat to an immediate superior will not harm the employee since reporting problems to supervisors is an ordinary part of the job. Therefore, employees are morally obligated to report serious threats to their immediate superiors. The employee must also report the problem to his or her superior so that the firm has an opportunity to correct the situation. This preserves the employee's loyalty to the employer. DeGeorge adds that if the superior already knows of the danger, then reporting it to him or her would be redundant, and the second condition would already be satisfied.

The third condition reads: "If one's immediate superior does nothing effective about the concern or complaint, the employee should exhaust the internal procedures and possibilities within the firm."[5] This condition also makes sure that employees do not violate their loyalty to their employers. Once the first three conditions are satisfied and the problem has still not been resolved, the employee has gained the right to override the demand of loyalty and blow the whistle on the organization. DeGeorge anticipates the most serious problem connected to this condition when he declares that if the harm is imminent, there may not be time to exhaust the internal procedures. He suggests that prudence and judgment should be used, but that at least some effort should be made in this direction.

In general, DeGeorge thinks whistleblowing is permissible when there is a threat of serious harm to the public and when the potential whistleblowers have done everything possible to inform their employers of their concerns. Whistleblowing would not be permissible without informing the employer since this would violate the employee's duty of loyalty. The permissibility of whistleblowing depends upon an action by the employees: making their concerns known within the organization; and a lack of effective action by the employer: failing to resolve the problem.

The fourth condition is the first of the two additional conditions that make whistleblowing ethically obligatory. "The whistleblower must have, or have accessible, documented evidence that would convince a reasonable, impartial observer that one's view of the situation is correct, and that the company's product or practice poses a serious and likely danger to the public or to the user of the product."[6] This condition raises doubts about whether anyone in a complex case would be ethically obligated to blow the whistle. First, it would be difficult for some potential whistleblowers to get access to such documentation since it might be closely guarded to protect it from competitors. It also might be of questionable legality for them to take the documentation if it is something that they are not entitled to have access to, or if it is proprietary information. Second, an even more serious difficulty is that it is unclear what it takes to convince a "reasonable, impartial observer." Reasonable people can view the same documentary evidence and not come to exactly the same conclusion, as the DC-10 case demonstrates. It is likely that potential whistleblowers would often be significantly uncertain about whether the documentation would convince a reasonable, impartial observer, and thus it would be rare for someone to be obligated to blow the whistle.

The fifth condition and the second of the two necessary to make whistleblowing ethically obligatory is: "The employee must have good reasons to believe that by going public the necessary changes will be brought about. The chance of being successful must be worth the risk one takes, and the danger to which one is exposed."[7] This last condition is also accompanied by difficulties. First, what counts as "good reasons" for believing that the changes will be brought about? Reasonable people might disagree about what constitutes "good reasons." It seems likely that "good reasons" would vary with the circumstances so there is no independent or general standard for this criterion. Furthermore, one such circumstance is that having "good reasons" is connected to knowing who will use the information and how it will be used, but it is often impossible to predict this with accuracy. DeGeorge's discussion provides no helpful guidance on this matter, and even if he tried, I do not think that he could come up with anything that would eliminate the inevitable uncertainty connected to the "good reasons" requirement. Once again, it seems that there would often be sufficient uncertainty to prevent ethical obligation. A second serious problem is connected to the question of how the employee is supposed to know if the chance for success makes the risk acceptable. The condition forces the employee to once again predict future events and to risk retaliation based on a prediction. DeGeorge does claim that the greater the risk to the employee, the greater the chance for success ought to be, but this is not very helpful by itself. There are no guidelines for comparing degree of risk to chances of success. As with the fourth condition, the fifth condition makes it unlikely that anyone would be obligated to blow the whistle. Potential whistleblowers might always be uncertain about whether the chances for success were worth the risk.

Conditions four and five are so demanding that they inject an unavoidable element of uncertainty into the evaluation of any complex case. Employees will almost never be obligated to blow the whistle on their employers in complicated cases. The problem in the fourth condition is the uncertainty connected to knowing what it would take to convince a reasonable, impartial observer that the employee's view of the situation is correct. The difficulty with the fifth condition is the uncertainty related to employees knowing if the changes will be brought about, and how should they compare the degree of risk to the chances for success. Instead of being conditions designed to determine whether the employee is ethically obligated or not, conditions four and five look more like requirements that prevent ethical obligation in almost all complex cases. The second part of this chapter will illustrate this state of affairs as DeGeorge's conditions are applied to the DC-10 case, in particular to the case of Dan Applegate, a high-level engineer involved in the development of the plane.

PART II

The DC-10 was a "jumbo jet" built by the McDonnell Douglas Corporation in the late sixties and early seventies. Douglas subcontracted most of the detail design of the fuselage, including the cargo doors, to the Convair Corporation. F. D. "Dan" Applegate was director of product engineering for Convair's part of the DC-10 project. On June 27, 1972, he wrote a memorandum to his immediate superior, J. B. Hurt, which identified a serious problem with the cargo doors and stated that he believed that if something was not done to improve the safety of the plane, sooner or later there would be a crash. Included in the memo was the statement, "It seems to me inevitable that, in the twenty years ahead of us, DC-10 cargo doors will come open and I would expect this to usually result in the loss of the airplane."[8] Applegate's prediction came true in the Paris crash of 1974 when 346 people were killed.

In the DC-10 case, Applegate was convinced that the situation represented "serious and considerable harm to the public." DeGeorge is concerned with cases that threaten serious harm or danger to life or health, and in the DC-10 case the problem with the cargo door did threaten the death of airline passengers and crew members. Therefore, the first condition would be satisfied.

Applegate reported the "serious threat" to his superior at Convair by sending J. B. Hurt his memorandum. Therefore, DeGeorge's second condition was satisfied also. It is harder to know whether the third condition was satisfied in Applegate's case. His memo was sent to J. B. Hurt, and Applegate and Hurt discussed the matter with M. C. Curtis, who was in charge of the DC-10 project for Convair. Since Curtis was at the top of the line for the DC-10 project, it might be suggested that Applegate had fulfilled the third condition. However, DeGeorge's condition states that the employee should ". . . exhaust the internal procedures and possibilities within the firm."[9] To really exhaust the possibilities within the firm, Applegate should have gone to Curtis's superior with his concerns. Thus, it might be argued that the third condition was not satisfied and it would not have been morally permissible for Applegate to blow the whistle on his employers. The resolution to the question of whether this condition is fulfilled seems to rest on DeGeorge's claim that if harm is imminent, there may not be time to exhaust the internal procedures and channels. DeGeorge believes that some effort should be made, but that the matter must rest on the employee's judgment. In this case, if Applegate felt that serious harm was imminent, he could have concluded that the third condition was satisfied because he had made some effort. Therefore, based on DeGeorge's first three conditions, it would have been ethically permissible for Applegate to blow the whistle in the DC-10 case.

With the examination of the first three conditions completed, it is important to see whether conditions four and five were satisfied. Although it would have been permissible for Applegate to blow the whistle, was he ethically obligated to do so? Condition four is tied to the issue of whether Applegate could have gotten ". . . documented evidence that would convince a reasonable, impartial observer" that his view of the situation was correct.[10] To answer this question, we must have a better idea of what constitutes Applegate's "view of the situation." Applegate's memorandum not only mentioned the problem with the plane, it also suggested how the deficiency might be corrected. He discussed recommending to Douglas the installation of blow-out panels in the cabin floor which would accommodate the explosive decompression without the loss of control of the aircraft. Clearly he thought this correction was the best possible answer. He referred to the post-Windsor installation of the inspection windows and the revising of the door-latch rigging as "more band-aids." Based on his experience, blowout panels were a reasonable response to the problem, but his evaluation of the situation differed from the opinions of the engineers at Douglas. Applegate's "view of the situation" is complex since it not only involves the problem, but also a particular way of fixing the problem.

Would Applegate have been able to get access to documentation that would have convinced a reasonable, impartial observer that his view of the situation was correct? It is reasonable to assume that as director of product engineering for Convair's part of the DC-10 project, Applegate had access to the test results on Ship One that first illustrated the explosive decompression problem. After July 6, 1972, he could have also gotten a copy of the National Transportation Safety Board report on the Windsor incident which would further document the nature of the problem. These documents would have convinced a reasonable observer that there was a problem with the plane that required fixing, but not necessarily that Applegate's solution to the problem was the only appropriate one. Since it is unclear whether convincing documentation was available, Applegate would not be ethically obligated to blow the whistle. This illustrates the earlier observation about the unavoidable element of uncertainty in complex cases and shows that even if condition five were satisfied, Applegate would not have been not ethically obligated to blow the whistle.

The final condition is: Would Applegate have good reasons to believe that by going public the necessary changes would be brought about? Would his chance of being successful be worth the risk he was taking, i.e. the loss of his job?[11] The judgment on condition five would be clearer if condition four had been satisfied. If the documentation had been adequate to convince a reasonable, impartial observer that Applegate's solution to the problem was the correct one, there would have been a good chance for condition five to be satisfied also. In the absence of convincing documentation, is there any chance for Applegate to have other good reasons to believe that he would be successful?

If Applegate had wanted to do as much as possible to get his solution to the plane's deficiency implemented, what should he have attempted to do? The most likely answer is that he should have contacted the FAA about his solution. After his July 5 meeting with Hurt and Curtis, Applegate should have known that Convair was not going to pressure Douglas to upgrade the safety of the DC-10. Convair was also not going to contact the FAA directly since it was forbidden by contract from doing so. Informing McDonnell Douglas directly of his concern would not have led to the correction he proposed either since Douglas had already rejected the solution he favored. It is important to reiterate that Applegate's memo declared that he was dissatisfied with the post-Windsor improvements ordered by Douglas. Therefore, it seems that his only hope for safety improvement lay with the FAA since they alone had the authority to force McDonnell Douglas to make the plane safer along the lines indicated by Applegate in his memo.[12]

If Applegate had wanted to contact the FAA, it would have been natural for him to contact the head of the Western Region, Arvin Basnight. The Western Region is where both Convair and Douglas are located. After the Windsor incident, Basnight and his staff were already at work on how to prevent another incident or a crash. On July 6, the NTSB made a formal recommendation to the FAA, which included modifying the DC-10 cargo doors and strengthening the cabin floor. Another one of the recommendations was the installation of blow-out or relief panels in the floor. The FAA, however, was looking for an intermediate fix which could keep the planes in service. The initial answer suggested by the Western Region seems to have been to issue an airworthiness directive which would have mandated beefed-up wiring, the installation of a support plate for the torque tube, and a peephole to check the position of the locking pins. They were overruled by the head of the FAA, John Shaffer, who made a "gentleman's agreement" with Jackson McGowen of Douglas that allowed Douglas to issue maintenance directives which would direct that the changes be made when the planes came in for regular maintenance. The planes remained in service and over the next two months Douglas sent out two service bulletins in addition to the one already issued about the improved wiring. One adopted the peephole idea, and the other recommended a support plate for the torque tube, which had bent in the Windsor incident, and an adjustment to the door linkages. These were all changes which Applegate's memo had argued were inadequate. (It seems that the support plate was not incorporated in Ship 29, and this and the misrigging of the door contributed to the Paris crash.)

Basnight and Shaffer would probably have disagreed with Applegate's solution. Would he have known this? It is difficult to say. If he had read the NTSB report on Windsor, he would have known that the investigators agreed with him. He might have believed that if he added his voice to theirs, there was a good chance that the blow-out panels might be mandated. The opposite conclusion is also possible however. If he had been aware that the FAA was looking for a fix which would keep the planes in service or had known that high officials at the FAA would be unsympathetic, he might have concluded that his warning would be ignored. He would have been aware that the Western Region had originally certified the plane and knew a great deal about it. They had also received the NTSB report on the Windsor incident, which made the same recommendation as he did. In spite of the NTSB report, they were only preparing a service bulletin which would require the improved wiring and the installation of a peephole in the door. Applegate's information would not have told them anything new, and they might have ignored him. Because of the vagueness of the condition and the complexity of the case, it is unclear whether Applegate would have had reason to believe that his blowing the whistle would bring about the changes he wanted.

The application of conditions four and five to the DC-10 case shows the difficulty in applying DeGeorge's conditions and illustrates the problems mentioned earlier. Cases where someone is ethically obligated to blow the whistle will be rare since it will usually be unclear whether the documentation (if he or she has been able to get it) would convince an impartial observer that the employee's view of the situation is correct or whether the employee has good reasons to believe that by going public the necessary changes would be brought about. For DeGeorge, it will sometimes be ethically permissible to blow the whistle, but rarely ethically obligatory. Whistleblowing becomes a matter where the employee must earn the right to blow the whistle by discovering a case of serious harm and giving the company the chance to correct the problem. This view-of whistleblowing protects individual employees, but in many cases risks the lives of a greater number of people in society at large. The 346 people who died in the Paris crash and their relatives would agree that a stronger account of ethical obligation and whistleblowing is necessary.

PART III

Are businesspeople and engineers ever ethically obligated to blow the whistle on their employers? The second approach to this question will center around a moral principle appropriate to the whistleblowing dilemma instead of appealing to a set of conditions. In his book *Practical Ethics*, Peter Singer proposes a principle that he thinks should be acceptable to most philosophers.[13] A slightly altered version of the principle is: if it is in our power to prevent something bad from happening without thereby sacrificing anything of comparable moral significance, then we ought to do it.[14] If this principle is accepted, it generates ethical obligations and can be used to create ethical obligations connected to whistleblowing. Legitimate whistleblowers attempt to prevent bad things from happening by informing someone of a problem with a product or practice. Questions arise about whistleblowing because other people disagree about the badness of the things in question, because they sometimes think that the whistleblowers are sacrificing something of comparable moral significance, and because potential whistleblowers are sometimes unclear about whether it is in their power to prevent the bad thing from happening.

The potential whistleblower's first step is to determine whether there is wrongdoing, i.e., is something bad occurring or about to occur? Second, the question about whether something of comparable moral significance is being sacrificed must be answered. Finally, is it in the whistleblower's power to prevent the wrongdoing? The principle states that if one can prevent something bad from happening without sacrificing anything of comparable moral significance, then one ought to do it. In a potential whistleblowing case, if one can prevent a wrongdoing and if this prevention is of more ethical significance than the other factors, then one is obligated to blow the whistle.

How does a whistleblower know if something "bad" has or is about to happen? It will not provide an adequate answer to simply say that "bad" translates into harmful. "Harmful" is insufficient since many products present the possibility of harm to their users. When someone does carpentry work, power tools are certainly potentially harmful, but the experienced carpenter knows the danger and acts accordingly. Cigarettes present a different case of harm. Unlike the case of power tools, the harm is unavoidable, yet the danger is well known by the public. It would be absurd for some tobacco company employee to blow the whistle on his employer about the fact that cigarette smoking causes cancer. There are also products which may cause harm, but the harm is insufficient to warrant whistleblowing. A tennis ball might harm someone in a minor way if it hit him or her. If the person were struck in the eye, the harm might be more serious, but the chances of this are remote. The harm associated with tennis balls is too minor or remote to warrant whistleblowing.

If "harmful" will not serve to identify "something bad," how can potential whistleblowers know if something bad has or is about to happen? One place to start might be to suggest that something "bad" at least includes things that cause death or serious physical harm, but this requires further explanation of "serious physical harm." It would either be arbitrary or impossible to create necessary and sufficient conditions for "serious physical harm," but some reasonable guidelines may be suggested. The purpose of the guidelines will be to identify paradigm cases, rather than providing a standard which would allow one to make a determination in all cases.

"Serious physical harm" should include, at least, disfigurement, sterilization, and injuries that require medical treatment to prevent loss of life.[15] Thus, the potential whistleblower should be able to distinguish to a large degree when a product is doing serious physical harm. There may also be a limited number of cases where this determination cannot be made, e.g., where we are not sure whether medical treatment was necessary to prevent loss of life.

The second aspect of the principle about "comparable moral significance" produces even greater problems than providing guidelines for what is "bad." How do potential whistleblowers know if they are sacrificing something of comparable moral significance when they try to prevent something bad from happening? The question might be answered by taking a consequentialist and humanist approach to moral significance. What is morally significant is what benefits and harms human beings, and is brought about by human agency. We are not sacrificing anything of comparable moral significance if we suffer a lesser harm to prevent a greater harm.

The comparison of benefits against harms, benefits against alternative benefits, and harms against alternative harms should take place on a case-by-case basis with certain ethical paradigms as guidelines. One simple example of such an ethical paradigm might be that the death of a person is worse than the loss of a job. If by blowing the whistle, someone could achieve the benefit of saving a life while suffering the harm of losing his or her job, the person ought to blow the whistle. The calculations will usually not be this easy since there may be a number of harms and benefits to be considered, but they will be possible in some cases, e.g., the DC-10 case. In part 4, we shall see how the calculation could be carried out in this case.

The third aspect of the principle suggests that we should prevent the bad thing from happening if it is in our power to do so. This determination must also be made on a case-by-case basis and there will be cases where the individual is unsure whether it is in his or her power to prevent the harm. What are we to say about these cases? Our usual reaction in straightforward events is to attempt to prevent a serious harm or a death if there is a chance to do so without incurring a greater or equal harm to ourselves. Even if I am not sure I can reach the drowning boy in time, I will attempt to save him. I send money to the most reputable famine relief organizations I can find, even though I am not certain that the money will reach those who are starving. The uncertainty does not prevent us from trying to prevent great harms. Therefore, in cases of uncertainty where one would not be sacrificing something of equal moral significance, it is the individual's obligation to try to prevent the harm if there is a reasonable possibility of doing so. This sets up a very demanding standard, but without it, it will usually be unclear whether someone is obligated to blow the whistle. There is uncertainty connected to most of the events in life, and if a degree of uncertainty releases us from ethical obligation, there will be very few cases where we are ethically obligated to do anything.

It might be suggested that the whistleblowing literature could be used to provide paradigm cases which would provide some guidance about the application of this principle to whistleblowing cases. The careful discussion of cases like A. H. Robbin's Dalkon Shield, Ford's Pinto of the 1970s, Johns-Manville's treatment of its asbestos workers, McDonnell Douglas's DC-10, and others might show how people were ethically obligated to blow the whistle in these cases because they had a reasonable chance of preventing something bad from happening, and the harm they would have prevented outweighed the harm their whistleblowing would have caused.

PART IV

The DC-10 case is a good example to use to try out this approach to whistleblowing. The principle which informs this position is: If it is in our power to prevent something bad from happening without thereby sacrificing anything of comparable moral significance, then we ought to do it. Did Applegate know that something "bad" was likely to happen? As we saw earlier, his memo clearly states that if the problem was not fixed, it would lead to the loss of a plane or planes and to the deaths of the passengers and crew. Therefore, it is clear that he knew there was the potential for something bad to happen.

Would Applegate have been sacrificing something of comparable harm by blowing the whistle and preventing the loss of hundreds of lives (assuming that his whistleblowing could have prevented the deaths)? This impales us upon one of the main problems with consequentialism: it forces us to compare different kinds of things. Would saving hundreds of lives, preventing the suffering of the relatives and friends of the victims, avoiding the bad publicity associated with the Paris crash, and saving the money paid to the victim's relatives in civil suits have outweighed the harm to Applegate and his family from his losing his job, the consequences of his violating his corporate loyalty, the inconvenience to thousands of airline passengers whose plans would be altered if the planes were grounded to install the blow-out panels, the cost to McDonnell Douglas and/or Convair of altering the planes, and the bad publicity associated with grounding the planes? As odd as it may be to make this comparison, to an ethical consequentialist the saving of hundreds of lives stands out as the most significant factor. If those lives could really have been saved, it would have outweighed the harms caused by the whistleblowing.

This brings us to the final and most difficult point: Was it in Applegate's power to get the plane fixed in the way he suggested by blowing the whistle on Convair and McDonnell Douglas? Did he have a chance of being successful? We have already seen that the most likely route for Applegate to take was to contact the FAA's Western Region office. Applegate's judgment would have confirmed the NTSB report's recommendation to install the blow-out or relief panels in the floor, but the FAA was looking for an intermediate fix which could keep the planes in service. Would the FAA have listened to him and mandated the change he wanted? Earlier it was concluded that it would have been impossible for him to know for sure whether he would have been successful. There is another factor which would have decreased his chances of success. It was not emphasized earlier because it was irrelevant to whether he would have thought he would be successful since he could not have known about it.

A serious problem for Applegate would have been that the likely section of the FAA for him to contact would have been the Western Region office and yet, these people were cut out of the decision by Shaffer. They wanted to issue an airworthiness directive, but Shaffer's gentleman's agreement overruled their plan. The deal was made before Basnight could get in touch with Shaffer, and thus Applegate's blowing the whistle to the Western Region would not have changed anything initially.

It is also possible that Applegate might have reached the FAA indirectly by going through the media or someone in Congress. If the media or an ambitious member of Congress had learned of his concerns, the matter might have been publicized and the FAA might have been pressured into issuing an airworthiness directive that incorporated the NTSB recommendations. While Applegate's word would have stood against the combined weight of Convair, McDonnell Douglas, and possibly some employees of the FAA, his inability to achieve immediate success did not mean that he would never be successful. As it happened, almost two years passed before the Paris crash, and he could have done a lot in that time. There would have been a chance for his claim that the "band-aid" repairs were inadequate to receive a full hearing, and thus a reasonable chance that he would have been successful.

The principle, slightly modified, states that if there is a reasonable chance that it is in our power to prevent something bad from happening without thereby sacrificing anything of comparable moral significance, then we ought to do it. Applegate believed that something bad would inevitably happen and he would not have sacrificed anything of comparable moral importance by blowing the whistle on Convair and McDonnell Douglas. The real problem with this case is whether it was in his power to prevent the Paris crash. This is a question which cannot be answered conclusively. It seems clear that he would not have been successful by simply working within Convair or contacting McDonnell Douglas. It also seems that he would not have been successful contacting the Western Region of the FAA. His best chance for success lay in blowing the whistle to a legislator or the media. It is possible that he had a chance to be successful by blowing the whistle and therefore, based on this demanding position, he was ethically obligated to do so. If he wanted to act ethically, he had to risk his job and blow the whistle.

In general, Dan Applegate's conduct was no worse and perhaps better than that of the other main players in this tragedy. At least Applegate was concerned enough about the matter to write his memo. Hurt and Curtis at Convair, McGowen of Douglas, and Basnight and Shaffer of the FAA all seem to have had more power to order the installation of the blow-out panels than Applegate, but presumably they did not believe that the installation of the panels was necessary to prevent a crash. Since the plane that crashed near Paris had only had some of the changes made to it and after the Paris crash the blow-out panels were installed on all DC-10 aircraft, it is impossible to say whether Applegate or the others assessed the situation correctly.[16] The DC-10 case was a complicated and tragic series of events. This chapter has focused on Dan Applegate's involvement in the case, but the conduct of these other people and of other Douglas and Convair engineers would bear scrutiny also. We need to hold engineers and businesspeople to high ethical standards if we want to prevent similar occurrences in the future. The view of whistleblowing and ethical obligation proposed in part 3 would do this.

When, if ever, are businesspeople or engineers ethically obligated to blow the whistle on their employers? The application of DeGeorge's conditions is one way to answer this question. Another answer is to say that if there is a reasonable possibility that it is in someone's power to prevent something bad from occurring by blowing the whistle and that by doing so he or she would not sacrifice anything of equal moral significance, then the individual is obligated to do it. I believe that DeGeorge's conditions make ethically obligatory whistleblowing too rare. The second approach to whistleblowing is preferable because uncertainty about documentation and whether the harmful situation will be corrected do not preclude the ethical obligation to blow the whistle. Businesspeople and engineers occupy positions where they can contribute to great human harm and benefit. We need to hold them to high standards, like those set up by the second view of whistleblowing and ethical obligation, so that there will be fewer tragedies like the Paris crash of 1974.

NOTES

1. Slightly different definitions of whistleblowing can be found in the following sources: Norman Bowie, *Business Ethics* (Englewood Cliffs, NJ: Prentice Hall, 1982) and Richard DeGeorge, *Business Ethics*, 3rd edition (New York: Macmillan Publishing Co., 1990).

2. Dan Applegate is the most prominent figure in the literature concerning the DC-10 case, and that is why I have chosen to focus on him. Although the focus is on Applegate, there were other engineers and corporate officials at Convair and McDonnell Douglas whose conduct in the case deserves scrutiny also.

3. DeGeorge, *Business Ethics*, p. 208.

4. Ibid., p. 210.

5. Ibid., p. 211.

6. Ibid., p. 212.

7. Ibid.

8. Paul Eddy, Elaine Potter, and Bruce Page, *Destination Disaster* (New York: New York Times Book Co., 1976), p. 258.

9. DeGeorge, *Business Ethics*, p. 211.

10. Ibid., p. 212.

11. It seems safe to assume that if Applegate had blown the whistle, he would have lost his job. To have his best chance at success, he would have had to blow the whistle openly. Since Convair was forbidden by contract to contact the FAA directly, they would have probably moved quickly to distance themselves from Applegate, and he would probably have been fired or suspended.

12. Applegate might also have contacted the media and urged them to make the matter public, hoping to put pressure on the FAA.

13. Peter Singer, *Practical Ethics* (Cambridge: Cambridge University Press, 1979).

14. Ibid., p. 168.

15. A more liberal approach might be to suggest that serious harm is done whenever one requires medical treatment.

16. Based on my position, it was unethical of Applegate to fail to blow the whistle on Convair and McDonnell Douglas. I am not claiming that it was unethical of McDonnell Douglas to fail to install the blow-out panels. This would be hard to establish since the plane that crashed had not had all the post-Windsor changes made to it and the door was misrigged. Perhaps the blow-out panels were not really necessary.

13

What is Hamlet to McDonnell Douglas or McDonnell Douglas to Hamlet?: DC-10*

On the third of March, 1974, on the outskirts of Paris, France, a Turkish Airlines plane carrying 346 passengers and crew fell from the sky killing all aboard. The plane was built by the McDonnell Douglas Corporation, with major design subcontracts to the Convair Division of General Dynamics. It was Ship 29 of the DC-10 line. It became clear shortly after the event that the crash was no mere accident or an act of God or due to pilot or crew error. Ship 29 fell from the sky when its cargo-hold door blew open at approximately 10,000 feet, causing the floor of the passenger compartment to collapse, thereby breaking the electrical and hydraulic lines that run under that floor. Without electrical or hydraulic power the airplane is unflyable.

The Paris crash was not the first cargo door failure on an in-

---

* Reprinted with editorial changes by permission of the author. Copyright ©
1982 by Peter A. French.

flight DC-10. Some two years earlier over Windsor, Ontario, an almost duplicate accident occurred, but not all hydraulic lines were severed. The pilot of that plane, owned by American Airlines, miraculously managed to land the airplane in Detroit, without fatalities. The history of DC-10 door failure and floor collapse, however, dates back to July, 1970 when Ship 1 of the line, while under pressurization tests outside a hangar at the Douglas factory, blew its cargo door and the floor collapsed.

There can be little doubt that many engineers and managerial personnel at McDonnell Douglas (and Convair) knew, well before the Paris crash, of the potential for a Class IV hazard[1] due to defective design of the DC-10 cargo door latching system and the floor structure. Not unexpectedly, McDonnell Douglas tried to blame Turkish Airlines and its ground personnel for the Paris disaster. When that did not stick, it then suggested that some relatively low-ranking members within its own corporate structure had been contributorily negligent in the manufacture of Ship 29. There is, indeed, strong evidence that some employees of McDonnell Douglas, through negligence, carelessness, or sloppiness, contributed to the design and manufacture of a defective airplane, but the authors of one of the books that provide the history of the corporate development of the DC-10 have also written:

> Some part of the blame (for the Paris disaster) must lie with the major subcontractor for the DC-10, the Convair division of General Dynamics. But the *central responsibility*, at least in terms of morality, must lie with McDonnell Douglas and in particular with its Douglas Division.[2]

Evidence supporting that claim will be of primary concern here, but the moral responsibility of some individuals within the broader context of corporate responsibility will also be examined. I am not concerned with legal responsibility, which is a matter for the courts. The standard legal notions of strict liability, enterprise liability, and, the traditional ambit of corporate legal responsibility, vicarious liability would seem to provide adequate legal bases for recovery for wrongful death suits in this case, but the usual interpretation of such notions steadfastly sidesteps full-fledged corporate *moral* accountability and blameworthiness and hence does not locate a ground upon which a viable theory of corporate criminality and punishment may be built. That project has been the basic intent of my work on corporate moral responsibility. The Paris disaster happens to provide particularly clear examples to test my theory[3] regarding corporate and individual moral responsibility against our intuitions.

In order to examine some of the morally significant aspects of this case, it will be necessary to introduce what I take to be rather commonly held views about the accountability of moral persons. A first condition of accountability (or moral responsibility) is what Bradley called "self-sameness." Bradley wrote:

> If when we say, "I did it," the I is not to be the one I, distinct from all other I's; or if the I now here is now the same I with the I whose act the deed was, then there can be no question whatever but that the ordinary notion of responsibility disappears.[4]

This condition is deeply grounded in our intuitions. It is, simply, the statement of the need for identity through time of moral subjects. If moral persons lack identity through time, in other words if there were *no* grounds for identifying one moral person now as the same person who existed at some previous time and did certain things, all sense of holding persons accountable for their deeds would be lost. The "self-sameness" condition, as I have argued elsewhere,[5] can be met by corporations as well as by natural persons.

In Book Three of the *Nicomachean Ethics*, Aristotle argues that only voluntary actions are properly praised or blamed. When behavior is involuntary, "we are pardoned and sometimes even pitied."[6] For Aristotle, behavior under constraint or compulsion or due to ignorance is involuntary. Let us concentrate on ignorance. Aristotle distinguishes two primary senses of ignorance. The first is ignorance of moral principle, which he tells us is not pardonable. He writes, "Ignorance in moral choice does not make an act involuntary—it makes it wicked."[7] Aristotle's second major sense of ignorance is "ignorance of the particulars which constitute the circumstances and the issues involved in the action."[8] Aristotle maintains that exculpability and pardon depend on what aspects of the circumstances are those of which the person is ignorant.

Consider Hamlet. Let us stipulate that he knows his moral principles, that he knows that intentionally killing an innocent human being is murder, is morally wrong. We also know that he is certainly ignorant of the fact that the person behind the arras in Gertrude's room is the rather harmless Polonius and not the murderous King Claudius. In fact, when he stabs that person, he believes he is stabbing the king. In keeping with Aristotle's point, it is clear that a person may be ignorant of many of the true descriptions of his action that are different from the one under which he intended the action. An action may be intended under one description but not intended under another. Hamlet intends to kill the person hiding behind the arras, and he intends to kill the King who he believes is hiding there, but he certainly does not intend to kill Polonius. Hamlet's ignorance of the identity of the person he intentionally stabs even may be offered in support of the claim that it is false that Hamlet *murdered* Polonius. (You can kill but you cannot unintentionally murder someone.) Of course, there is much more we should want to say about Hamlet's deed before we consider letting him off the moral hook. At the very least, however, it is clear that when accountability is at issue, it is crucial to know what a person intended as well as to know what actually happened. Insofar as ignorance limits what a person can actually intend, we see why true descriptions of a person's actions that are unknown or unforeseeable by him are not properly praised or blamed. Suppose Hamlet had stabbed into the arras thinking he was killing a rat and, lo and behold, the king, who had been hiding there, is dispatched on his way off this mortal coil. Would Hamlet deserve praise for so efficiently avenging his father? Not very likely!

These considerations may be captured in what I have elsewhere called the strict or Primary Principle of Accountability (PPA):[9] a person can only be held accountable for that person's intentional acts.

PPA, however, is counterintuitive. It doesn't even satisfy in the Polonius stabbing case. A few minor modifications need to be made.

First, we do want to hold persons morally accountable, at least to some degree, for some of the unintended effects of their actions, those they should have or did know would occur. Suppose you know that you have inadequate skill for the performing of some task that when performed by the unskilled usually results in harm to someone else. The first time you perform the task, if you are unaware that harm will result and harm does result, we are likely to write off the harm as an accident or a misfortune and not hold you accountable. If you perform the task again, without sufficiently improving your skill, even if causing the harm is not your intention, it would generally be regarded as true that you were willing to have the harm occur. In such cases you can be held morally accountable for the harm. This explains why we are inclined to hold Hamlet morally accountable for Polonius' death. When he stabs through the arras with the intention of killing the person behind it (though he believes that it is the king), he is willing to kill Polonius if, as it happens,

Polonius is the person in hiding. Polonius cries out from behind the arras, "Oh, I am slain!" The queen asks Hamlet: "What hast thou done?" And Hamlet, somewhat befuddled, responds: "Nay, I know not. Is it the King?" When he learns it is Polonius, however, he says without apparent remorse: "Thou wretched, rash, intruding fool, farewell! I took thee for thy better." (His lack of regret at having killed Polonius shows again when he carts off the corpse.) It seems not unfair to describe Hamlet as having at least been willing to kill Polonius, though he certainly had no intention of doing so. Being willing to do something then does not entail intending to do it, but that means that moral accountability involves more than PPA allows. Not having intended the outcome, of course, might be treated as mitigatory in some cases. Holding Hamlet morally responsible for Polonius' death is not necessarily equivalent to accusing him of murder.

The second modification of PPA takes into account obliquely or collaterally intended second or non-original effects that involve other persons. Suppose we have two persons: John and Mary. Suppose John does a and Mary does b at some time later, but in direct response to John's doing a, and an outcome of Mary's doing b is harmful, and John was aware of that when he did a. Should John or Mary or both be held accountable for that harm? If Mary's doing b is a natural or (within some organizational structure) a required response to John's doing a, we usually hold only John primarily accountable for the harm (or John and the organization) by reason of what we may call oblique intention of a non-original effect. (We assume that John knows that to get Mary to do b, he has to do a or that he can get Mary to do b if he does a.) If Mary's doing b, although a response to John's doing a, is something she knows to be morally wrong and she can be truly described as willing for the harmful event to occur, even if not intending to cause harm, then Mary and not John may be held primarily accountable for the harm, *ceteris paribus*. But, if John's doing a is a clear temptation for Mary to do b, and if John should know or does know it is such a temptation and John does a, (even if he does not intend that Mary do b) and harm occurs, then the moral responsibility for the harm would, to some extent or in some degree be distributed to both, but fall more heavily on John than Mary, *ceteris paribus*. Within certain organizational structures, however, Mary's doing b may be an established and automatic response to John's doing a, and that should be or is known by John. In such cases, when harm results, John is held more to account and Mary less so for the result of Mary's doing b. The historically

more automatic Mary's response, the less she is held morally to
account.

The principle of accountability that satisfies all of these considerations could be put in the following form: A person (pre-sup-
posing Bradleyian self-sameness) may be held morally accountable
for his intentional actions and for those actions that he was willing to
perform under different descriptions of his intentional actions. Also,
he may be held accountable for those non-original or second effects
that involve the actions of other persons that he obliquely or collaterally intended or was willing to have occur as the result or under
different descriptions of his actions. Let us call this the Extended
Principle of Accountability (EPA). Armed with EPA, which may still
need further modification but will suffice for present purposes, let us
examine the DC-10 case. Remember that our first interest is with the
ascription of moral responsibility to McDonnell Douglas.

What might justify such as ascription? It seems to me that we
have at least three solid contenders: (1) McDonnell Douglas would
have to have decided to build Ship 29 intending that its design be
defective such as would predictably result in a crash that kills hundreds of people; or (2) McDonnell Douglas, not intending Ship 29 be
defective or crash, would have to have taken steps in the development, design, and construction process of the airplane that it knew
or should have known to be inadequate with regard to safety and
highly likely to result in an in-flight Class IV hazard; or (3) it would
have to have established policies and performed actions that it knew
or should have known would prompt rather automatic responses by
persons associated with the corporation that would increase the likelihood of the manufacture of a defective product, Ship 29.

Clearly McDonnell Douglas did not design, manufacture, and
sell DC-10s with the intention that they crash. It surely did not
build Ship 29 and sell it to Turkish Airlines with the intention of
killing 346 people. Our interest will have to focus on the other two
possibilities, one that puts forth the claim that without intending to
produce a defective airplane, McDonnell Douglas was willing to do so
and the other that claims that McDonnell Douglas was willing to
have a harmful outcome occur as the result of the predictable actions
of other persons made in response to actions or policies of McDonnell
Douglas. In both claims appeal is made to EPA and not to the strict
PPA. Surprisingly and sadly, the facts support both claims.

Before we examine those facts, it will help if we are clear about
a few rudiments of DC-10 design.[10] We need concern ourselves only
with cargo doors and passenger floors. The cargo doors on a DC-10

are large and cannot utilize the same plug design that secures the passenger entry doors. Instead, a latching system needs to be installed. There are two alternative types of systems that might be used. One is electrical, the other hydraulic. An electric latch actuator system is lighter, has fewer parts, and is easier to maintain than a hydraulic system. A hydraulic system, however, continuously exerts pressure on the latching device, holding it in place. An electrical system exerts pressure only when it is switched on. Electrically driven latches are prone to slip back if they are not made irreversible. When they achieve maximum force they must be fixed until the electric switch is again activated.

An electric latch will behave very differently from a hydraulic one. A hydraulic latch, though positive, is *not* irreversible. If it fails to go over-center, it will in the nature of things "stall" at a point where the pressure inside the cylinder has reached equilibrium with the friction which is obstructing the travel of the latch. Thereafter, quite a small opposite pressure will move it in the reverse direction—and what this means in a pressure-hull door is that if the latches have not gone quite "over," they will slide open quite smoothly as soon as a little pressure develops inside the hull and starts pushing at the inside of the door. Thus, they will slide back and the door will open, well before the pressure inside the aircraft hull is high enough to cause a dangerous decompression. The door will undoubtedly be ripped from its hinges by the force of the slip stream but, at low altitude, that poses no threat of structural damage to the plane and no danger to its passengers. The crew will immediately become aware of the problem, because the aircraft cannot be pressurized, and can simply return to the airport.

However, if an irreversible electric latch fails to go over-center, the result will usually be quite different. Once current is switched off, the attitude of the latch is fixed, and if it has gone quite a long way over the spool, there will be considerable frictional forces between the two metal surfaces, holding the latch in place. Pressure building up inside the door cannot *slide* the latches open. It can only force the fixed, part-closed latches off their spools. This, typically, will happen in a swift and violent movement, occurring only when pressure inside the airplane has built up to a level when sudden depressurization will be structurally dangerous.[11]

The underside of the passenger floor of a wide-bodied jet-liner is laced with the electrical and hydraulic lines that are absolutely necessary for the airplane to fly. The stability of that floor is crucial. In the event of a sudden decompressurization of the cargo area, enormous unoffset pressures from the passenger compartment are exerted on the floor. Without sufficient support, it will buckle and break and in so doing sever the electrical and hydraulic lines. The DC-10 was designed both with an electric cargo door latching system and with relatively few passenger floor supports, given the wide-bodied nature of the craft. Why? Certainly not because McDonnell Douglas intended to build a dangerous airplane.

McDonnell Douglas seems to have believed that by using technology that had proven successful during its long tradition of building passenger aircraft, it was manufacturing a safe product. DC-10 engineers, under express management orders, utilized the existing Douglas technology gained on the DC-3, DC-8, and DC-9 as the basic design for the wide-bodied DC-10.[12] That technology, though not clearly inappropriate to the jumbo craft, in the case of crucial systems, had been superceded by engineering advancements pioneered by Boeing and Lockheed on the 747 and the TriStar. The "state of the art" had developed, before McDonnell Douglas had committed itself to its DC-10 design, beyond the design constraints with which McDonnell Douglas had saddled its engineers, and the relevant technology was not proprietary.[13]

McDonnell Douglas has an oft-stated[14] company policy of technological caution that, combined with its severe financial straits in the 1960s, was apparently interpreted by its engineers and manufacturing staff to dictate that corners be cut[15] and existing Douglas technology be used, even if that meant that some systems that were rejected as inferior by its competitors would be designed into the DC-10. It is of note that another engineering result of policy and financial constraint is less redundancy of key systems in the DC-10 than on 747s and TriStars. Both its competitors have four hydraulic systems each of which is capable of providing sufficient power for a landing. (Manual power is ineffective in moving control surfaces on a wide-bodied airplane.) The DC-10 was engineered with only three redundant hydraulic systems, all running in parallel under the cabin floor. Its competitors can weather the loss of one more system than the DC-10. There is, of course, a point somewhere along the line where redundancy, even in essential systems, is waste. It probably is not reached, however, at three or four systems.[16] The DC-10 simply did not achieve the minimum state of the relevant engineering art.

In the case of some products, that might be only quaint, but not, we should all allow, when the transportation of millions of people is involved.

The evidence supports the claim that McDonnell Douglas designed an airplane that they *should have known* did not meet the engineering standards of the industry with respect to certain crucial systems. Manufacturing that airplane is arguably redescribable as being willing to produce a Class IV in-flight hazard, regardless of what McDonnell Douglas might have intended. This sounds extremely harsh; unfortunately for McDonnell Douglas matters are worse. Not only *should they have known* of the extreme hazard to life they were manufacturing, they *did* know. The evidence seems incontrovertible.[17] As it happens, in order to make the DC-10 case analogous to Hamlet's killing Polonius, we would have to imagine that Hamlet in fact knows that the chances are the person hiding behind the arras is not the king, yet he stabs away intending to kill the king, but willing to kill whoever is there, perhaps because he wants, at least, to terrify his mother.

In 1969 Convair was asked by McDonnell Douglas to prepare a Failure Mode and Effects Analysis (an FMEA) for the DC-10's cargo door latching system. The FMEA is an assessment by the design engineers of the likelihood of failure of the system and its consequences for the airplane. FMEAs on the critical systems of a new airplane must be given to the FAA by the manufacturer when the airplane is certified. Convair's FMEA for the cargo door latching system shows that there were at least nine possible failure sequences that could result in life-endangering hazards. Four of those sequences would produce sudden depressurization in flight and the almost certainty of a crash of the airplane. One of those sequences in the FMEA reads as follows:

> Door will close and latch, but will not safety lock. Indicator light will indicate normal position. Door will open in flight—resulting in sudden depressurization and possibly structural failure of floor; also damage to empennage by expelled cargo and/or detached door. Class IV hazard in flight.[18]

Approximately five years later that is exactly what happened over the outskirts of Paris. (It may also be of note that McDonnell Douglas never submitted that Convair-prepared FMEA to the FAA.[19]) If the FMEA was not enough evidence for McDonnell Douglas to be said to *know* of the defects of its DC-10 design, one year later, in

1970, the cargo door blew and the floor collapsed on Ship 1, the prototype of the line, while it was undergoing standard pressurization tests.

If we apply EPA to the facts, it is clear that on the basis of this evidence along McDonnell Douglas can be held morally accountable for the Paris crash. Its knowledge of the defective Class IV hazard design of the cargo door latching system well before Ship 29 was built for Turkish Airlines provides us with adequate grounds to support the claim that McDonnell Douglas was willing to manufacture and market an airplane that had a higher probability than the Boeing 747 or the Lockheed L-1011 of creating a Class IV hazard in flight. Again notice that I do not believe McDonnell Douglas wanted or intended to manufacture a defective airplane. Its intentional actions regarding design decision, however, clearly have a high probability, as shown in its own FEMAs, of certain consequences which, though unintended, ought, nonetheless, given the information available to McDonnell Douglas, to have been expected by the manufacturer. Sadly, there is even evidence, in the form of company memos, that high-ranking engineering personnel expected Class IV hazards to occur on in-flight DC-10s.[20] It is as if Hamlet did not expect it was the king in hiding, hoped it was, and stabbed anyway.

In this analysis I have not tried to locate within the McDonnell Douglas corporate structure any individual human beings who were significant contributors to the series of corporate decisions that resulted in the design and manufacture of the DC-10. Finding the corporation morally accountable surely ought not to exculpate such persons. In cases such as this, however, the corporate "black box"[21] is not easily penetrated. The effects of what have been called the "Law of Diminishing Control" and "Cognitive Dissonance" as well as unintentional blocking of information, etc., in such large decentralized firms as McDonnell Douglas, belie any simple reductionism to the believes, reasons, and intentions of human beings associated in the corporation. Australian corporate legal theorist W. B. Fisse has (with a nod to my work) recently written:

> The conventional assumption has been that corporate crime reduces to the willed acts of individual actors. However, this assumption fails to account for corporate behavior which cannot be explained exclusively in individualistic terms . . . corporate policy is not merely the sum of individual intentions but a collective choice influenced and constrained by organizational factors, including bargaining and teamwork. Nor are corporate

acts simply the aggregation of individual acts: organization is a *sine qua non*. Accordingly, there is no oddity about regarding a corporation as a criminally responsible actor (or a moral person) where the act alleged has a sufficient organizational nexus.[22]

There is, however, another aspect of the unhappy history of Ship 29 in which both individual and corporate responsibility, in accord with EPA, may be assessed.

After McDonnell Douglas admitted possible difficulties with locking the cargo doors following the Windsor incident, an agreement with the FAA was reached that called for a modification of the doors. In July, 1972, Ship 29 was in the "Rework for Delivery" area of the Long Beach plant of McDonnell Douglas. Included in the work to be done on Ship 29 were modifications to the cargo door. The plant records for July, 1972, indicate that three inspectors stamped the work records for Ship 29 to indicate the modifications had been completed and that the plane was in compliance with FAA guidelines. None of the work on the cargo doors had actually been done. All three inspectors identified the stamps on the work record as theirs, but none of the three remembered having worked on Ship 29. There is no evidence that the stamps of all three, however, were stolen or borrowed during July, 1972. It is company policy that:

> You are responsible for any work that your stamp appears on the record for accepting.[23]

One inspector, Edward Evans, conjectured that his stamp appears on the records of Ship 29 either because "it was high summer,"[24] that he had become confused between airplanes because of the summer heat and had stamped the wrong document, *or* that he had been interested in some other aspect of the reworking and had not carefully inspected the cargo door latching system. The other inspectors could not even offer conjectures as to how their stamps appeared on the records of Ship 29.

The president of the Douglas division of McDonnell Douglas, John Brizendine, who was responsible for "engineering, flight development, and production" of the DC-10, when questioned by attorneys, claimed no personal knowledge of misuse of inspection stamps. In fact, he insisted that he had nothing to do with insuring that design reworking actually was done. After the Paris crash he did reprimand the inspectors for misuse of stamps.

We could give two alternative accounts of what happened with regard to the inspections of Ship 29. The first is that three inspectors, whether singly or in concert, lied or conspired to lie about the inspections of Ship 29. But for what reason? The second is that the McDonnell Douglas system that is supposed to assure that individual responsibility is exercised by personnel at various stages of production to "insure quality in the tradition of the company" is fundamentally weak and easily compromised by employees who have fallen into a rather automatic pattern of behavior encouraged by that company policy and procedure. It is certainly impractical to insist that President Brizendine or any other high-ranking corporate office make all of the necessary inspections, etc., himself. However, it was not responsible for him to assume an uncompromised verification of inspection procedure.

The evidence[25] supports the view that over the years McDonnell Douglas established an inspection procedure that invites or tempts inspectors to be lax and careless and some of those inspectors, either through inadvertence or because of conditioning to laxness, cursorily performed tasks that, given the basically poor design of the aircraft, called for the closest attention to detail to insure safety. McDonnell Douglas policies and procedures constituted a temptation to carelessness, even though managerial superiors were unaware of such a temptation and that safety was at stake. Those corporate officers, we may say with some confidence, should have expected that their inspection procedures would be unintentionally compromised by inspectors.

The actions of the three inspectors are not excusable (if inadvertence is proved, that may be exculpatory), but it would be a grand offense to our moral intuitions, in the absence of any evidence of intentional sabotage, etc., to hold those inspectors primarily responsible for the crash of Ship 29. We are brought back to the principal actor in the design, manufacture, and sale of Ship 29, McDonnell Douglas Corporation.

When recently discussing this DC-10 case with certain economists and management professors, I was surprised to learn that they thought the whole matter should be resolved as a risk issue. On their account, McDonnell Douglas, operating in the best of market traditions, simply had taken a not unreasonable, calculated risk in the manufacture of the DC-10. All machines are liable to failure or breakdown, and any manufacturer of a machine needs to take the probability of failure into account when marketing the product. For example, it buys product liability insurance. Also, any consumer of the machinery, I was told, can be presumed to know that machines are liable to breakdown. Turkish Airlines and its passengers can be presumed to have had such general knowledge of machinery. What then is the problem? If all that is meant by "taking a risk" in a business context is that the corporation, fully cognizant of certain product deficiencies, produces and markets the product in the hope (probably supported by statistically based predictions of failure) that sales will offset any actual liabilities due to inadequate design, then evidence that such calculated risks were taken can only serve to strengthen the ascription of moral responsibility to the corporation on the occasion of harm. "It was a calculated risk" has no exculpatory or even mitigatory power. In fact, in cases of this sort, if anything, it is inculpating. There is, however, another side to the risk issue that is often confused with the matter of whether the corporation took a reasonable or calculated risk when it decided to manufacture and market its product. That is the matter of putting persons using the product "at risk" in cases where ordinarily they would be presumed to have taken or accepted the risk.

Clearly, risk, in this sense, is, in most cases, negotiable. A reasonable person will accept a certain amount of risk if the compensation is satisfactory. Everyone, we might say, has a risk budget. But risk cannot be negotiated in a unilateral contract nor can it be said to have been negotiated in the absence of knowledge by the parties of the relevant data. For risk to have been negotiated, the facts relevant to the potential for the causing of harm must be known by the parties, at least in general terms. Certainly every passenger who enters a commercial airliner should realize that there is a certain probability that the aircraft will develop a malfunction that could result in his death. In the case of most American commercial airplanes that probability is remarkably low. The benefits of fast travel, etc., far outweigh the risks in the minds of many travelers. The passenger pays for his ticket and boards the airplane. In effect, there is a certain, generally understood, risk to life involved in the use of a commercial airliner. Passengers can be presumed to understand that the risk is about the same for any craft on which they might fly.

None of this rather unexceptional business is, however, relevant to the DC-10 case. The DC-10 design was such as to drive it out of the risk probabilities of its sister craft. Risk budget calculations made in ignorance of the faulty design are not necessarily going to be the same as those that would have been made if the facts were known. The relevant design information was, of course, not widely known or easily accessible to consumers. In the case of Ship 29, Turkish Airlines clearly should have known of the previous cargo door locking failures, but it also had the assurances of McDonnell Douglas that FAA-required modifications had been made.

Risk is simply not a unilateral matter. McDonnell Douglas's decision to take a risk with the DC-10, of course, supports the view that is uncontested, that McDonnell Douglas had no intention of building a defective airliner. The fact is McDonnell Douglas was just wrong in its calculation of the risk it was taking with respect to product liability. It was creating a far greater risk for the passengers of its aircraft than they could have taken into account when deciding to fly the DC-10, and it should have known (indeed it did know) that it had miscalculated the risk. If the information about the defective design of the DC-10 had been public knowledge, it is unlikely McDonnell Douglas would have been able to sell the airplane. (After the Chicago crash of 1978, attributed to engine mount failure, the entire DC-10 fleet was grounded for reworking and inspection. The resumption of DC-10 service, despite bargain rates and other allurements [compensations], was greeted with less than enthusiasm by the flying public. Even today many passengers report a preference to book flights on planes other than DC-10s. That is, of course, a good example of risk bargaining at work in the marketplace. No such circumstances surround the Paris crash.)

In short, the introduction of the notion of risk in the DC-10 case is totally irrelevant to the moral accountability concern. Even if all 346 people who boarded Ship 29 in Paris on March 3, 1974, knew they were entering a machine and that every machine has a statistical probability of breakdown for any moment of its operation, they cannot reasonably be said to have struck a bargain to accept the realistic risk probabilities of the *crash of Ship 29* for the compensation of the convenience of less in-travel time between Paris and London. McDonnell Douglas' failure to engineer to the state of the art again enters the picture of our analysis because it accounts for the fact that the probabilities of harm generally understood to apply to American commercial aircraft were inapplicable to the calculation of the risk passengers were taking in the case of DC-10s. The facts are that McDonnell Douglas knowingly exposed DC-10 passengers to a significantly higher probability of death than they would have been exposed to on other aircraft. The fact that McDonnell Douglas performed acts that constituted creating such a risk and concealed pertinent evidence from the public and the regulatory agency only supports the moral indictment.

In the absence of admission of corporations like McDonnell Douglas to citizenry in the moral world, the moral responsibility for the Paris air disaster cannot reasonably be finally assessed. The aggregate of justifiable individual responsibilities for the production of Ship 29 simply does not "add-up to" that for its crash. Without a theory of the corporation as a moral person upon which to base the accountability ascriptions I have made in this analysis, the real villain of the piece will escape moral detection. A. A. Berle has written:

> The medieval feudal power system set the "lords spiritual" over and against the "lords temporal." These were the men of learning and of the church who, in theory, were able to say to the greatest power in the world: "You have committed a sin; therefore either you are excommunicated or you must mend your ways." The "lords temporal" could reply: "I can kill you." But the "lords spiritual" could retort: "Yes that you can, but you cannot change the philosophical fact." In a sense this is the great lacuna in the economic power system today.[26]

My theory of the corporation as a moral person[27] uncovers the grounds necessary to redescribe the actions of persons associated with a corporation as intentional acts of that corporation. If my account is cogent, we will have found the missing link that brings the corporate giants, today's "lords temporal," into the scope of morality.[28]

NOTES

1. A Class IV hazard is a hazard involving danger to life.

2. Paul Eddy, Elaine Potter, and Bruce Page, *Destination Disaster* (New York: New York Times Book Co., 1976).

3. See my "The Corporation as a Moral Person," *American Philosophical Quarterly*, July, 1979.

4. F. H. Bradley, *Ethical Studies* (Oxford: Oxford University Press, 1876, 1970 ed.), p. 5.

5. "Crowds and Corporations," *American Philosophical Quarterly* (forthcoming).

6. Aristotle, *Nicomachean Ethics*, Ostwald translation (Indianapolis: Bobbs-Merrill Company, Inc., 1962), p. 52.

7. Ibid., p. 55.

8. Ibid., pp. 55-56.

9. *The Scope of Morality* (Minneapolis: University of Minnesota Press, 1979), chapter 1.

10. For a complete account see Eddy, Potter, and Page, *op. cit.*

11. Ibid., pp. 176-177.

12. Ibid., chapter 6, especially pp. 96-99.

13. Ibid., pp. 85-99 provide a clear and adequate account of the state of the art at the time of the development of the DC-10.

14. The authors of *Destination Disaster* quote from McDonnell Douglas company literature references to Pope's "Be not the first by whom the new are tried/Nor yet the last to lay the old aside" and Carnegie's "Pioneering don't pay." Further accounts of corporate policy are cited in chapter 6.

15. Ibid., p. 97.

16. Ibid., p. 99.

17. Ibid., especially chapter 10.

18. Ibid., p. 178.

19. Ibid., pp. 176-177.

20. Ibid., chapter 10.

21. See Christopher Stone, *Where the Law Ends* (New York: Harper and Row, 1975.

22. W. B. Fisse, "The Retributive Punishment of Corporations" (1980), unpublished.

23. Eddy, Potter, and Page, *Destination Disaster*, p. 223.

24. Ibid., p. 224.

25. Ibid., pp. 223-235.

26. A. A. Berle, "Economic Power and the Free Society," *The Corporate Take-Over*, ed. by A. Hacker (Garden City, Doubleday, 1964), p. 99.

27. See note 3.

28. The author acknowledges with appreciation support from the Exxon Education Foundation and the Center for the Study of Values, University of Delaware.

Commentary*

In his article "What is Hamlet to McDonnell Douglas or McDonnell Douglas to Hamlet?: DC-10," Peter French stresses that, in his opinion, the McDonnell Douglas Corporation should have been held morally accountable for the crash of a DC-10 commercial jetliner that caused the death of 346 people in March of 1974. I heartily agree with his basic thesis. Also, I would allow as self-evident the fundamental tenets of moral philosophy which French calls to our attention—that a morally accountable individual must have been in some way the causal agent of the morally significant action in question and that moral culpability presupposes intention in some degree. I have some difficulties, however, with French's specific treatment of the situation regarding the DC-10 safety deficiencies.

There are a number of significant considerations concerning moral accountability for these deficiencies. Collectively, designers and producers of artifacts for general public use have a fundamental moral obligation to design and produce such articles to meet safety standards that are reasonably compatible with the potential of the article to cause injury, as a result of malfunction when being used pursuant to design intent, to users and/or others who may be coincidentally within range of effect. The principle is applicable whether the article is a safety pin, an airplane, or a hydroelectric dam. In cases where the potential for cause of injury generally implies severe injury and/or death for substantial numbers of people, that obligation becomes a mandate to design products to the maximum standards of safety technologically feasible at the time. If this entails costs that make public use of the artifact too expensive, the public will simply reject use of the article unless or until technological progress makes possible the reduction of cost without compromise of safety stan-

---

* Reprinted with editorial changes by permission of the author. Copyright © 1982 by Homer Sewell.

dards. The designers and producers, however, are never at liberty with respect to moral accountability to compromise safety standards as a means of lowering production costs.

No misunderstanding exists among managerial and technical personnel in the aerospace industry concerning the fundamental moral obligations of companies with respect to product safety. Any supposition to the contrary is simply naiveté with respect to the real-life substratum of knowledge, awareness, and ethics that characterizes the industry. Compromise of that obligation is by definition either inadvertent (employees of aerospace firms are no more immune to carelessness than other human beings) or intentional, and intentional compromise in the interest of cost reduction and/or delivery schedule commitments is not without precedent.

As the documentation cited in French's article establishes, McDonnell Douglas has been rightfully subject to question on four counts related to safety in connection with design and production of the DC-10. The initial design of the cargo door latching and locking mechanism was patently deficient with respect to safety standards reasonably achievable through application of available technology. The state of deficiency unambiguously known to exist by a substantial number of McDonnell Douglas's engineering and engineering/management employees, arose out of morally unjustifiable subversion of safety considerations to cost considerations. If the deficiency had been deliberately allowed to persist on aircraft delivered to airline operators, moral accountability for ensuing injury and deaths to passengers and operating crew would be, prima facie, a burden that the knowing employees would have to bear.

Steps were taken, however, to correct the defective design prior to delivery of aircraft to airline operators. For reasons that have not yet been established, the necessary retrofix was not incorporated into an aircraft (tail no. 29) delivered to the Turkish Airlines which subsequently crashed. Quality control inspection stamps indicated that all measures normally taken to ensure incorporation of properly approved design changes into aircraft destined for delivery to airline operators apparently had been taken for aircraft no. 29. The three quality control inspectors involved insist that they have no idea how the inspection documentation could have acquired inspection stamps indicating incorporation of required retrofit design fixes when the fixes had not in fact been incorporated. The inspectors remember nothing decisive concerning the inspection or lack of inspection of aircraft no. 29. There are only three possible explanations for such a state of affairs: (a) the failure to inspect and detect the lack of retrofix incorporation was an inadvertent happening attributable to carelessness on the part of the inspectors, or (b) the inspectors deliberately stamped the production and test documentation without making the required inspection and have subsequently lied about their actions, or (c) other parties gained access to the inspection stamps and used them, in absence of any knowledge by the authorized inspectors.

In the case of (a), had it obtained, neither the inspectors in question nor any other employees of McDonnell Douglas could be held morally accountable pursuant to failure of the cargo door latching and locking mechanism for the ultimate crash of the aircraft and attendant loss of life. No one intentionally allowed a state of affairs to materialize which constituted a more than minimally obtainable safety hazard.

In the case of (b), had it obtained, the inspectors, each as sovereign morally accountable agents—which is, of course, what we all are with respect to morally significant actions—would be directly morally responsible for the resulting deaths.

It cannot be stressed too strongly in this regard that quality control (QC) inspectors in airplane manufacturing companies, both in this country and elsewhere in the world where airplanes are being manufactured for use by commercial airlines, know to the very depth of their innermost beings that the QC inspection stamp is a sacrosanct confirmation that the configuration of the product, as it moves from stage to stage during manufacturing and testing, is either precisely what the engineering documentation (drawings and specifications) requires or that any deviation therefrom is positively identified in the configuration accounting records and has been authorized by legitimate engineering authority.

It is not the QC inspector's duty to second-guess as to whether engineering has required adequate safety conditions; that is the job of engineering supervision. It is, however, supremely and exclusively his job to verify that manufacturing has faithfully produced what engineering has specified. In that capacity he constitutes the first line of defense for produce safety integrity—and he knows it.

In the case of (c), had it obtained, ascription of moral accountability would be a function of the extent to which the "others" in question were aware of the required retrofit change to the cargo door latching and locking mechanism configuration and the intrinsic relationship between that requirement and the operational safety index for the aircraft. It is possible to postulate both scenarios wherein such awareness could be zero, and a basis for ascribing moral accountability for the crash and consequent loss of life accordingly nil, or the awareness would be comparable to a QC inspector's awareness—suppose the other(s) was/were other QC inspectors!—and moral culpability would be correspondingly absolute. We will, in all probability, however, never know any more about the QC inspection hiatus on the DC-10 no. 29 than we know today—and will, accordingly, never be able to more than theorize about how it may relate to ascription of moral accountability vis-à-vis the crash of no. 29.

I do not agree with French that the evidence at hand suggests a need for drastic overhaul of McDonnell Douglas's QC inspection policies and procedures. An objective comparison of those requirements with their counterparts at Boeing, General Dynamics, Lockheed, Grumman, British Aerospace, etc., would not reveal significant differences. The policies and procedures employed are in no way proprietary to a particular company; they have evolved from over half a century of exponentially expanded experience and have become incredibly effective from a statistical point of view. Carelessness, human cupidity, and innocent acts of both commission and omission cannot be completely eliminated from endeavors in which human beings play a significant role. A comparison of aircraft inflight safety degradation due to QC inspection error or laxness with the degradation that stems from pilot error, however, would reveal the former to be essentially insignificant.

Electrical and hydraulic subsystems on the DC-10 include only two degrees of back-up redundancy to provide the fail-safe failure modes, whereas the same subsystem on the Boeing 747 and on the Lockheed L-1011 include three degrees of redundancy back-up. It is easy (and probably equally superficial) to infer from that criterion that DC-10s are inherently less safe than 747s and L-1011s. It would be virtually impossible to prove it, however, in an empirical sense. There is good statistical support for first degree redundant back-up for many subsystems on aircraft. The cause-effect relationship between failure of those subsystems and crash is too direct and inexorable, and the margin between minor crashes and crashes that result in serious injury and/or loss of life is too narrow to permit exclusive reliance on primary subsystem performance.

On commercial airliners carrying hundreds of passengers, the case for a second back-up subsystem is made plausible by the sheer number of people at risk. At that stage, however, we are involved more with engineering philosophy than with engineering judgment based on empirical evidence. When we start considering the need for still another—a third—back-up subsystem, we have nothing in the way of criteria but "but feeling" to appeal to. There is no honest objective verification that three or more degrees of redundancy significantly increase the safety index of the overall situation. There are competent, conscientious engineers who will maintain that the third redundancy is justifiable obeisance to safety conservatism. There are other equally competent and conscientious engineers who will call incorporation of third degree redundancy unmitigated cost irresponsibility.

More pertinent to the issue at hand, however, is the fact that in the area where electrical and hydraulic lines run beneath the floor that separates the passenger from the cargo compartments on large commercial airliners, no degree of subsystem redundancy will provide fail-safe protection against catastrophic floor rupture. If the floor goes, the hydraulics and electrics go also. Therefore, it is essentially specious to accuse McDonnell Douglas of safety irresponsibility because of failure to incorporate third degree redundancy for the DC-10 electrical and hydraulic subsystems.

The design and construction of the DC-10 floor is another matter, however. It is technically possible to equip the aircraft with a floor that will withstand rapid decompression of the cargo hold at operating altitudes without buckling or fracturing and severing electrical and hydraulic lines underneath. It is also possible to place—as in he 747—the electrical and hydraulic lines above the passenger compartment ceiling where they are in no way vulnerable to passenger compartment floor failure. Adopting initially either a beefed-up and properly vented floor or placing electrical and hydraulic lines overhead would have increased the overall cost of the DC-10 airframe. If that increment of cost would have made sale of the airplane impossible at a price that could successfully compete against the 747 and L-1011 while providing an acceptable profit margin for the manufacturer, then McDonnell Douglas should not have proceeded with the sale, production, and delivery of the airplane. Under no circumstances should they have delivered to commercial airlines an aircraft that they knew to have an unsafe floor design and unsafe location of vital electric and hydraulic lines. As a matter of fact, the cost issue must have been marginal at most, because McDonnell Douglas did not withdraw the DC-10 from the market when they were forced by the Federal Aviation Administration after the crash of no. 29 to substantially improve the floor design and to retrofit that safety deficiency fix into all of the DC-10s that had already been delivered to airline customers.

I believe that the engineers/managers at McDonnell Douglas who were aware of the safety implications of the original floor configuration should have been charged in the DC-10 no. 29 case with criminal negligence and prosecuted to the full extent of the law. They are, in my opinion, guilty, prima facie, of manslaughter, at the least.

It is nonsense to assert that, in an organization as large as the McDonnell Douglas Corporation, it is impossible to distinguish the responsible parties from their fellow employees. In any aerospace company, a clear hierarchy of responsibility for basic product-design decisions begins with first-level engineering supervision and would, in the case at issue, reach up to and include, as a minimum the individual executive in charge of the design engineering function in the company.

It is not so long ago, historically speaking, that the principal manufacturers of electrical apparatus in this country got together and colluded in the fixing of prices for things like large electrical motors and generators, transformers, and switchgear, etc., in violation of the Sherman Antitrust laws.[1] To the consternation of virtually all of U.S. industry, the government refused to treat the matter as simple violation of civil law by "amorphous corporate entities," which would have led at worst only to stiff fines for the corporations involved. Instead, it charged a number of officials of those companies with criminal conduct and imposed jail sentences along the personal fines. It seems to me that if violation of the Sherman Antitrust laws is serious enough to merit criminal prosecution, there should be no question about how society should proceed against individuals who have exposed the public to conditions which fall flagrantly short of reasonable safety standards.

NOTE

1. See John G. Fuller, *The Gentleman Conspirators* (New York: Grove Press, 1962). The original indictments were handed down in May of 1959. The major companies involved were: The General Electric Company, Westinghouse, Allis Chalmers, Federal Electric, and International Telephone and Telegraph.

14

Statement of John C. Brizendine, President, Douglas Aircraft Company, McDonnell Douglas Corporation

Good morning, gentlemen. My name is John C. Brizendine and I am president of Douglas Aircraft Company, a division of McDonnell Douglas Corporation. With me is Ray E. Bates, vice president-Design Engineering.

It is the policy of our company to fully cooperate with this committee [House of Representatives Special Subcommittee on Investigations of the Committee on Interstate and Foreign Commerce] and to provide all information and documents requested. We welcome this opportunity to present our company position and to state our complete and total confidence in the safety and airworthiness of our family of Douglas commercial aircraft including the DC-10 aircraft now operated by twenty-three airlines throughout the world. At the outset, it should be pointed out that the investigation of the Turkish Airlines DC-10 accident near Paris on March 3, 1974 has not been completed and no conclusions have been reached regarding the cause of that accident. While it does appear from preliminary information that the initiation of the accident may have been the separation of the aft cargo door from the aircraft, it is incumbent upon the investigatory authorities in Paris to completely investigate all possible causes of the accident in the interest of public safety. It is also incumbent upon our company and the cognizant United States authorities to insist that this be done.

It has been reported from Paris that a key part of one of the modifications to the cargo door, which I will discuss later, was found missing upon inspection of the aft cargo door located at the accident site. According to our manufacturing records, all service bulletins that had been issued to improve the cargo door latching mechanism had been incorporated in the Turkish Airlines aircraft prior to its delivery. At this time we are unable to satisfactorily explain this discrepancy, but we are continuing to investigate this matter. The same discrepancy was also found as to another aircraft and the part has since been installed on that aircraft. It must be remembered that the door on the Turkish DC-10 would have been perfectly secure if it had been properly latched and locked. As soon as Douglas learned of the possible involvement of the aft cargo door in the Turkish accident, we took immediate action on March 4, 1974 and notified all DC-10 operators, by telegram, of that possibility and strongly recommended enforcement of previously issued cargo door latching instructions. We also arranged meetings with both domestic and foreign operators of the DC-10 aircraft (March 7, 1974 with U.S. carriers with resident representatives of foreign carriers also in attendance, and March 11, 1974 with foreign carriers in Geneva, Switzerland) in order to establish a program for the accelerated incorporation of the most recent improvements to the cargo door latching mechanism then in production and for which service bulletins had previously been issued. The Air Transport Association and Douglas arranged a joint meeting in Long Beach on March 7, 1974, attended by representatives of Douglas, ATA, the U.S. operators of the DC-10, and FAA personnel from the Western Region and Washington, D.C., for the purpose of making this program known to the FAA and to commit our intentions. This program is now the subject of an FAA airworthiness directive and is well under way to a prompt completion.

In order to fully understand the design improvements to the cargo door system, it is appropriate to briefly review the history of the cargo door system on the DC-10 aircraft. The basic design concept of the latching and locking system used on DC-10 cargo doors has been on outward opening doors for Douglas pressurized aircraft since the DC-6 in the early 1950s. The latch consists of a heavy linkage-driven hook mounted on the door which engages a cylindrical spool mounted on a fixed structure in the door jamb. Upon engagement of the hook and spool, the linkage articulates to an "overcenter" geometry, the kinematics of which make it physically impossible for the hook to become disengaged from the spool. Further, the kinematics of the linkage are such that cabin pressure loads on the door are in the direction that maintain engagement of the hook and spool. For redundancy and double safety, lock pins are inserted behind the latch hooks to provide a physical barrier to hold the latch hooks in the closed position. The lock pins cannot be inserted to the locked position unless the hook is in the latched, overcenter position. An indicating system is provided, with annunciating lights in the cockpit, to advise the flight crew that each door is closed and latch lock pins are properly positioned.

Multiple latches are installed on each door (three to seven latches per door on the DC-10, depending on the size of each door) and the strength of the latches are such that any one of the latches can structurally fail without imparing the integrity of the door. Additional loss of latches is permissible if the latches are not adjacent.

On May 29, 1970, while conducting a ground pressurization test during the construction stage of DC-10 fuselage No. 1, the forward cargo door had been improperly latched and the door came open due to cabin pressure loads. The cabin floor sustained local damage in the vicinity of the door. Because of this incident, all cargo doors were redesigned to incorporate a vent door linked to the lock pin actuating handle and sized so as to make it impossible to pressurize the aircraft if the door was not properly latched and locked. This system was installed on all DC-10 aircraft prior to delivery.

On June 12, 1972, an aft cargo door separated from an American Airlines DC-10 while department from Detroit. Investigation of that incident revealed that the aft cargo door had not been properly latched and locked and that it had been possible for the mechanic to force the vent door closed, by forcing the handle down with a force over 120 pounds, without the lock pins being properly positioned. On June 16, 1972 Douglas issued an alert wire which was followed on June 19 by alert service bulletin A52-35 which installed a viewing port in each cargo door whereby the proper positioning of the lock pins could physically be observed from outside the aircraft. On June 19, 1972, the FAA issued a telegram to all DC-10 operators which reflected Alert Service Bulletin A52-35 and which had the same effect as a formal airworthiness directive. (At that time there were thirty-five DC-10s in service, operated by four airlines, all U.S. carriers.) In addition, Douglas designed further improvements to the cargo door latching system. For example, Service Bulletin 52-37 issued July 3, 1972, revised the rigging of the latch pins on all cargo doors and installed a support plate on the aft cargo door linkage which increased the handle force required to close the vent door when the locking pins were not engaged from a force of over 120 pounds to over 440 pounds. Douglas also continued design activity to further improve the cargo door latching system working with the airlines and the FAA in establishing workable criteria. This design activity has resulted in the concept known as the "closed loop system" which is now being incorporated in all DC-10 aircraft.

The DC-10 has been designed and manufactured to meet all Federal Aviation Regulation requirements. In fact, FAR criteria are considered by Douglas to be *minimum* standards and our practices go well beyond FAR in many instances. For example, FAR Part 25 requires fuselage structures to sustain a fail-safe load after failure of any single structural element such as one fuselage skin panel *or* one stiffener, whereas the DC-10 is actually designed to sustain fail-safe loads of multiple structural elements, specifically, two adjacent fuselage panels *plus* the stiffener. These capabilities are verified by component and full scale fail-safe testing.

The complete DC-10 airframe structural life was verified by full scale fatigue testing the equivalent of 120,000 flight hours and 84,000 individual flights, including full pressurization loads to the relief pressure of 9.1 psi on each flight cycle. There is no specific FAR requirement for fatigue life testing on fail safe structure, but our company performed the test which, incidentally, cost approximately $30 million.

For emergency landing conditions, the DC-10 aft engine restraint is designed for 12G forward combined with 6G downward compared to the FAR 9G forward requirement. There are many other examples which time today does not permit us to describe.

On the subject of floor venting, the DC-10 design provides for cargo compartment blowout areas ranging from 430 square inches in the aft bulk compartment to 970 square inches in the enlarged forward compartment without damage to the passenger floor. No discrete FAR requirements exists on this subject for certification, but approximately 130 square inches has been used historically. (This is based upon the general size of fuselage structural panels in a fail safe fuselage shell.)

There have been questions raised regarding the venting and strengthening of jumbo jet floors. Douglas Aircraft Company has conducted studies regarding floor venting and strengthening, along with several other design studies, and has discussed and reviewed these studies with interested parties. Douglas' primary efforts have been to insure the integrity of the pressure vessel, that is, the fuselage, doors and door latching mechanisms.

In addition to the large degree of difficulty of making such changes to delivered aircraft, there are significant practical problems with schemes to increase floor venting which could render such changes ineffective. For example, loose baggage could block the vents and reduce or eliminate their effectiveness.

Douglas Aircraft Company has recommended that the FAA consider the questions of venting and cabin floor strength as matters for fundamental review of design criteria under FAR's applicable to the design of all wide-bodied aircraft. Our own design studies are continuing and we pledge full cooperation with the FAA, other regulatory agencies and the industry at large.

I appreciate the opportunity to speak here today and I will be pleased to answer any questions you gentlemen may have.

Thank you.

THE 1979 CHICAGO CRASH

15

National Transportation Safety Board
Report on the 1979 Chicago Crash*

NATIONAL TRANSPORTATION SAFETY BOARD
WASHINGTON, D.C. 20594

AIRCRAFT ACCIDENT REPORT

Adopted: December 21, 1979

AMERICAN AIRLINES, INC.
DC-10-10, N110AA
CHICAGO-O'HARE INTERNATIONAL AIRPORT
CHICAGO, ILLINOIS
MAY 25, 1979

---

* Reprinted, with editorial changes, from *National Transportation Safety Board Report, NTSB-AAR-79-17* (1979).

SYNOPSIS

About 1504 c.d.t., May 25, 1979, American Airlines, Inc., Flight 191, a McDonnell-Douglas DC-10-10 aircraft, crashed into an open field just short of a trailer park about 4,600 feet northwest of the departure end of runway 32R at Chicago-O'Hare International Airport, Illinois.

Flight 191 was taking off from runway 32R. The weather was clear and the visibility was 15 miles. During the takeoff rotation, the left engine and pylon assembly [see fig. 16.1] and about 3 feet of the leading edge of the left wing separated from the aircraft and fell to the runway. Flight 191 continued to climb to about 325 feet above the ground and then began to roll to the left. The aircraft continued to roll to the left until the wings were past the vertical position, and during the roll, the aircraft's nose pitched down below the horizon.

Flight 191 crashed into the open field and the wreckage scattered into an adjacent trailer park. The aircraft was destroyed in the crash and subsequent fire. Two hundred and seventy-one persons on board Flight 191 were killed; two persons on the ground were killed, and two others were injured. An old aircraft hangar, several automobiles, and a mobile home were destroyed.

The National Transportation Safety Board determines that the probable cause of this accident was the asymmetrical stall and the ensuing roll of the aircraft because of the uncommanded retraction of the left wing outboard leading edge slats [see fig. 16.2] and the loss of stall warning and slat disagreement indication systems resulting from maintenance-induced damage leading to the separation of the No. 1 engine and pylon assembly at a critical point during takeoff. The separation resulted from damage by improper maintenance procedures which led to failure of the pylon structure.

Contributing to the cause of the accident were the vulnerability of the design of the pylon attach points to maintenance damage; the vulnerability of the design of the leading edge slat system to the damage which produced asymmetry; deficiencies in Federal Aviation Administration surveillance and reporting systems which failed to detect and prevent the use of improper maintenance procedures; deficiencies in the practices and communications among the operators, the manufacturer, and the FAA which failed to determine and disseminate the particulars regarding previous maintenance damage incidents; and the intolerance of prescribed operational procedures to this unique emergency.

The facts developed during the investigation disclosed that the

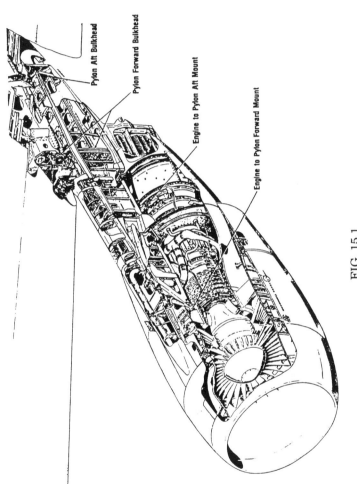

Pylon Aft Bulkhead

Pylon Forward Bulkhead

Engine to Pylon Aft Mount

Engine to Pylon Forward Mount

FIG. 15.1
Engine and Pylon Assembly

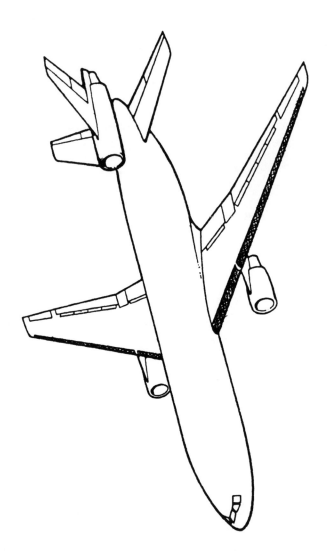

FIG. 15.2
DC-10 Slats

initial event in the accident sequence was the structural separation of the number one engine and pylon assembly from the aircraft's left wing. Witness accounts, flight data recorder parameters, and the distribution of the major structural elements of the aircraft following the accident provided indisputable evidence that the engine and pylon assembly separated either at or immediately after rotation and about the same time the aircraft became airborne. At that time, the flight crew was committed to take off, and their decision not to attempt to discontinue takeoff was in accordance with prescribed procedures and was logical and proper in light of information available to them.

The investigation and analysis were concentrated primarily in two major areas. First, the investigation sought to identify the structural failure which led to the engine-pylon separation and to determine its cause; second, the investigation attempted to determine the effects the structural failure had on the aircraft's performance and essential systems, and the operational difficulties which led to the loss of control. In addition, the investigation went beyond these primary areas and probed such areas as the vulnerability of the DC-10's design to maintenance damage, the adequacy of the DC-10's systems to cope with unique emergencies, the quality control exercised during DC-10 manufacturing and aircraft assembly, the adequacy of operator maintenance practices, the adequacy of industry communications of service and maintenance difficulties, the extent of FAA's surveillance of overall industry practices, and the adequacy of an accepted operational procedure.

About eight weeks before the accident, the No. 1 pylon and engine had been separated from the wing of the accident aircraft in order to replace the spherical bearings in compliance with McDonnell Douglas' service bulletins 54-48 and 54-59. The four other American Airlines and two Continental Airlines aircraft, in which cracks-were detected in the aft bulkhead's upper flange, had also been subjected to the same programmed maintenance during which the engine and pylon was removed. Further corroboration that the cracks had been produced during these maintenance operations was obtained when it was learned that Continental Airlines had, on two occasions before the accident, damaged the upper flange on the aft bulkhead as pylons were being removed or reinstalled. In these two instances, the damage was detected; the bulkheads were removed and repaired in accordance with a method approved by McDonnell Douglas.

Therefore, the evidence indicated that the overstress cracks in the aft bulkhead's upper flange were being introduced during a maintenance operation used by American and Continental Airlines. Both operators had devised special programs to replace the forward and aft bulkhead's spherical bearings. The manufacturer's service bulletins recommended that the maintenance be performed during an engine removal and that the engine be removed from the pylon before the pylon was removed from the wing. Both American Airlines and Continental Airlines believed that it would be more practical to comply with the service bulletin when an aircraft was scheduled for major maintenance—maintenance which would not necessarily otherwise necessitate engine removal. Therefore, American and Continental devised a procedure which they believed to be more efficient than that recommended by McDonnell Douglas—removal of the engine and pylon as a single unit. An engine stand and cradle were affixed to the engine and the entire weight of the engine and pylon, engine stand, and cradle was supported by a forklift positioned at the proper c.g. for the entire unit. The pylon-to-wing attaching hardware was removed, and the entire assembly was lowered for access to the spherical bearings. These were replaced and the entire unit was then raised and the attaching hardware reinstalled.

A close examination of these maintenance procedures disclosed numerous possibilities for the upper flange of the aft bulkhead, or more specifically the bolts attaching the spar web to this flange, to be brought into contact with the wing-mounted clevis and a fracture-producing load applied during or after removal of the attaching hardware in the aft bulkhead's fitting. Because of the close fit between the pylon-to-wing attachments and the minimal clearance between the structural elements, maintenance personnel had to be extraordinarily cautious while they detached and attached the pylon. A minor mistake by the forklift operator while adjusting the load could easily damage the aft bulkhead and its upper flange. The flange could be damaged in an even more insidious manner; the forks could move imperceptibly as a result of either an internal or external pressure leak within the forklift's hydraulic system during pylon removal. The testimony of the mechanics who performed the maintenance on the accident aircraft confirmed that the procedure was difficult.

The number one engine and pylon assembly separated after the flightcrew was committed to continuing the takeoff. Witnesses saw the pylon and engine assembly travel up and over the left wing after it separated, and the deformation of the pylon's forward bulkhead was consistent with their observations. The left wing's leading edge skin forward of the pylon's front bulkhead was found on the runway with the pylon structure. There was no evidence that the pylon and engine assembly struck any critical aerodynamic surfaces of the aircraft or any of the flight control surfaces.

Since the loss of thrust provided by the number one engine and the asymmetric drag caused by the leading edge damage would not normally cause loss of control of the aircraft, the safety board sought to determine the effects the structural separation had on the aircraft's flight control systems, hydraulic systems, electrical systems, flight instrumentation and warning systems, and the effect, if any, that their disablement had on the pilot's ability to control the aircraft.

The severing of the hydraulic lines in the leading edge of the left wing could have resulted in the eventual loss of number three hydraulic system because of fluid depletion. However, even at the most rapid rate of leakage possible, the system would have operated throughout the flight. The extended No. 3 spoiler panel on the right wing, which was operated by the number three hydraulic system, confirmed that this hydraulic system was operating. Since two of the three hydraulic systems were operative, the Safety Board concludes that, except for the number two and number four spoiler panels on both wings which were powered by the number one hydraulic systems, all flight controls were operating. Therefore, except for the significant effect that the severing of the number three hydraulic system's lines had on the left leading edge slat system, the fluid leak did not play a role in the accident.

During takeoff, as with any normal takeoff, the leading edge slats were extended to provide increased aerodynamic lift on the wings [see fig. 16.3]. When the slats are extended and the control valve is pulled, hydraulic fluid is trapped in the actuating cylinder and operating lines. The incompressibility of this fluid reacts against any external air loads and holds the slats extended. This is the only lock provided by the design. Thus, when the lines were severed and the trapped hydraulic fluid was lost, air loads forced the left outboard slats to retract. While other failures were not critical, the uncommanded movement of these leading edge slats had a profound effect on the aerodynamic performance and controllability of the aircraft. With the left outboard slats retracted and all others extended, the lift of the left wing was reduced and the airspeed at which that wing would stall was increased. The simulator tests showed that even with the loss of the number two and number four spoilers, sufficient lateral control was available from the ailerons and other spoilers to offset the asymmetric lift caused by left slat retraction at air-

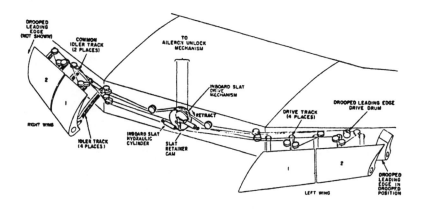

SYSTEM SCHEMATIC

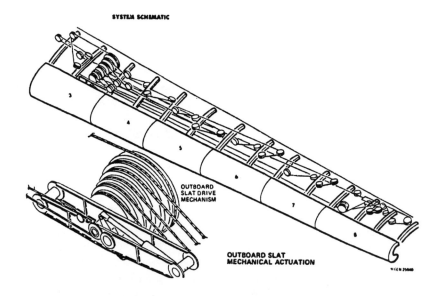

OUTBOARD SLAT
MECHANICAL ACTUATION

FIG. 15.3
Slat Mechanical Actuation

speeds above that at which the wing would stall. However, the stall speed for the left wing increased to 159 KIAS [knots indicated air speed].

The evidence was conclusive that the aircraft was being flown in accordance with the carrier's prescribed engine failure procedures. The consistent 14° pitch attitude indicated that the flight director command bars were being used for pitch attitude guidance and, since the captain's flight director was inoperative, confirmed the fact that the first officer was flying the aircraft. Since the wing and engine cannot be seen from the cockpit and the slat position indicating system was inoperative, there would have been no indication to the flight crew of the slat retraction and its subsequent performance penalty. Therefore, the first officer continued to comply with carrier procedures and maintained the commanded pitch attitude; the flight director command bars dictated pitch attitudes which decelerated the aircraft toward V_2, and at V_2 + 6, 159 KIAS, the roll to the left began.

The aircraft configuration was such that there was little or no warning of the stall onset. The inboard slats were extended, and therefore, the flow separation from the stall would be limited to the outboard segment of the left wing and would not be felt by the left horizontal stabilizer. There would be little or no buffet. The flight data recorder also indicated that there was some turbulence, which could have masked any aerodynamic buffeting. Since the roll to the left began at V_2 + 6 and since the pilots were aware that V_2 was well above the aircraft's stall speed, they probably did not suspect that the roll to the left indicated a stall. In fact, the roll probably confused them, especially since the stickshaker had not activated.

The roll to the left was followed by a rapid change of heading, indicating that the aircraft had begun to yaw to the left. The left yaw—which began at a 4° left wing down roll and at 159 KIAS—continued until impact. The abruptness of the roll and yaw indicated that lateral and directional control was lost almost simultaneous with the onset of the stall on the outboard section of the left wing.

The simulator tests showed that the aircraft could have been flown successfully at speeds above 159 KIAS, or if the roll onset was recognized as a stall, the nose could have been lowered, and the aircraft accelerated out of the stall regime. However, the stall warning system, which provided a warning based on the 159 KIAS stall speed, was functioning on the successful simulator flights. Although several pilots were able to recover control of the aircraft after the roll began, these pilots were all aware of the circumstances of the accident. All participating pilots agreed that based upon the accident circumstances and the lack of available warning systems, it was not reasonable to expect the pilots of Flight 191 either to have recognized the beginning of the roll as a stall or to recover from the roll. The safety board concurs.

The safety board is also concerned that the designs of the flight control, hydraulic, and electrical systems in the DC-10 aircraft were such that all were affected by the pylon separation to the extent that the crew was unable to ascertain the measures needed to maintain control of the aircraft.

Also, the influence on aircraft control of the combined failure of the hydraulic and electrical systems was not considered. When aircraft controllability was first evaluated based on asymmetric leading edge-devices, it was presumed that other flight controls would be operable and that slat disagree and stall warning devices would be functioning. Flight 191 had accelerated to an airspeed at which an ample stall margin existed. Postaccident simulator tests showed that, if the airspeed had been maintained, control could have been retained regardless of the multiple failures of the slat control, or loss of the engine and numbers one and three hydraulic systems. On this basis alone, the Safety Board would view the design of the leading edge slat system as satisfactory. However, the additional loss of those systems designed to alert the pilot to the need to maintain airspeed was most critical. The stall warning system lacked redundancy; there was only one stickshaker motor; and the left and right stall warning computers did not receive crossover information from the applicable slat position sensors on opposite sides of the aircraft. The accident aircraft's stall warning system failed to operate because d.c. power was not available to the stickshaker motor. Even had d.c. power been available to the stickshaker motor, the system would not have provided a warning based on the slats retracted stall speed schedule, because the computer receiving position information from the left outboard slat was inoperative due to the loss of power on the No. 1 generator bus. Had power been restored to that bus, the system would have provided a warning based on the slat retracted stall speed. However, in view of the critical nature of the stall warning system, additional redundancy should have been provided in the design.

In summary, the certification of the DC-10 was carried out in accordance with the rules in effect at the time. The premises applied to satisfy the rules were in accordance with then accepted engineering and aeronautical knowledge and standards. However, in retrospect, the regulations may have been inadequate in that they did not require the manufacturer to account for multiple malfunctions resulting from a single failure, even though that failure was considered to be extremely improbable. McDonnell Douglas considered the structural failure of the pylon and engine to be of the same magnitude as a structural failure of a horizontal stabilizer or a wing. It was an unacceptable occurrence, and therefore, like the wing and horizontal stabilizer, the pylon structure was designed to meet and exceed all the foreseeable loads for the life of the aircraft. Therefore, just as it did not analyze the effect the loss of a wing or horizontal stabilizer would have on the aircraft's systems, McDonnell Douglas did not perform an analysis based on the loss of the pylon and engine.

Logic supports the decision not to analyze the loss of the wing and horizontal stabilizer. With the loss of either of these structures, further flight is aerodynamically impossible and the subsequent effect of the loss on the aircraft's systems is academic. However, similar logic fails to support the decision not to analyze the structural failure and loss of the engine and pylon, since the aircraft would be aerodynamically capable of continued flight. The possibility of pylon failure, while remote, was not impossible. Pylons had failed. Therefore, fault analyses should have been conducted to consider the possible trajectories of the failed pylon, the possibilities of damage to aircraft structure, and the effects on the pilot's ability to maintain controlled flight. Since the capability of continued flight was highly probable, the fault analysis might have indicated additional steps or methods which could have been taken to protect those systems essential to continued flight.

Therefore, the Safety Board concludes that the design and interrelationship of the essential systems as they were affected by the structural loss of the pylon contributed to this accident.

Maintenance Programs

Although the Safety Board believes that the design of the pylon structure was less than optimum with regard to maintainability, the evidence is conclusive that many pylons were removed from the wing and reinstalled without imposing damage to the structure. There is no doubt, however, that this maintenance operation requires caution and extreme precision because of the minimal clearances at the pylon-to-wing attachment points and the danger of inadvertent impact of the structure.

McDonnell Douglas was apparently aware of the precision

which would be required, and as a result it specified in its original maintenance procedures and subsequent service bulletins that the engine be separated from the pylon before the pylon is removed from the wing. While removal of the engine would not completely eliminate the possibility of imposing damage to the pylon structure, the likelihood would certainly be much less than that which existed when handling the pylon and engine as single unit. The pylon assembly without the engine weighs about 1,865 lbs and the c.g. is located approximately three feet forward of the forward bulkhead attachment points. The pylon and engine together weigh about 13,477 lbs. and the center of gravity is located about nine feet forward of the forward bulkhead attachment points. With the engine removed, the pylon can be supported relatively close to the pylon-to-wing attachment points where precise relative motion between the pylon and wing structure can be closely observed and controlled. Thus, McDonnell Douglas did not encourage removing the engine and pylon assembly as a single unit because of the risk involved in remating the combined assembly to the wing attach points. The Safety Board, therefore, is concerned with the manner in which the procedures used to comply with Service Bulletins 54-48 and 54-59 were evaluated, established, and carried out.

American Airlines is a designated alteration station, as are the other major carriers that conduct heavy maintenance programs. Pursuant to that designation and the applicable regulations, carriers are authorized to conduct major maintenance in accordance with the maintenance and inspection program established by the FAA's Maintenance Review Board when the aircraft was introduced into service. Carriers are also authorized to conduct alterations and repairs in accordance with the procedures set forth in its maintenance manuals or established by its engineering departments. The FAA, through its principal maintenance inspectors, is responsible for surveillance of carriers' maintenance programs. However, this surveillance is broadly directed toward insuring that the carriers comply with the established maintenance and inspection program and that their maintenance programs, including administration, general practices, and personnel qualifications, are consistent with practices acceptable to the administrator. The FAA can review the carriers' maintenance manuals, but its formal approval is not required. Carriers are permitted to develop their own step-by-step maintenance procedures for a specific task without obtaining the approval of either the manufacturer of the aircraft or the FAA. It is not unusual for a carrier to develop procedures which deviate from those specified by the manufacturer if its engineering-and maintenance personnel believe that the task can be accomplished more efficiently by using an alternate method.

Thus, in what they perceived to be in the interest of efficiency, safety, and economy, three major carriers developed procedures to comply with the changes required in service bulletins 54-48 and 54-59 by removing the engine and pylon assembly as a single unit. One carrier apparently developed an alternate procedure which was used without incident. However, both American Airlines and Continental Airlines employed a procedure which damaged a critical structural member of the aircraft. The procedure, developed by American Airlines and issued under ECO R-2693, was within American Airlines' authority, and approval or review was neither sought nor required from the manufacturer or the FAA.

The evidence indicated that American Airlines' engineering and maintenance personnel implemented the procedure without a thorough evaluation to insure that it could be conducted without difficulty and without the risk of damaging the pylon structure. The safety board believes that a close examination of the procedure might have disclosed difficulties that would have concerned the engineering staff. In order to remove the load from the forward and aft bulkhead's spherical joints simultaneously, the lifting forks had to be placed precisely to insure that the load distribution on each fork was such that the resultant forklift load was exactly beneath the center of gravity of the engine and pylon assembly. To accomplish this, the forklift operator had to control the horizontal, vertical, and tilt movements with extreme precision. The failure of the ECO to emphasize the precision this operation required indicates that engineering personnel did not consider either the degree of difficulty involved or the consequences of placing the lift improperly. Forklift operators apparently did not receive instruction on the necessity for precision, and the maintenance and engineering staff apparently did not conduct an adequate evaluation of the forklift to ascertain that it was capable of providing the required precision.

The safety board, therefore, concludes that there were other deficiencies within the American Airlines maintenance program, some of which contributed to this accident. Among these was the failure of the engineering department to ascertain the damage-inducing potential of a procedure which deviated from the manufacturer's recommended procedure, their failure to adequately evaluate the performance and condition of the forklift to assure its capability for the task, the absence of communications between maintenance personnel and engineers regarding difficulties encountered and the procedural changes which were required in the performance of the pylon maintenance, and the failure to establish an adequate inspection program to detect maintenance-imposed damage. Although the safety board directed its investigation to American Airlines, the safety board is concerned that these shortcomings were not unique to that carrier. Since two of Continental Airlines DC-10s were found to have been flying with damaged bulkheads, similar shortcomings were also present in its maintenance program.

Industry Communications Regarding Maintenance Difficulties

The safety board is particularly concerned that because of the limitations of the current reporting system the FAA and key engineering and maintenance personnel at American Airlines were not aware that Continental Airlines had damaged two aft bulkhead flanges on two of its DC-10s until after the accident. In December 1978, after it discovered the first damaged bulkhead, Continental apparently conducted a cursory investigation and determined that the damage resulted from a maintenance error. A repair was designed for the bulkhead and was submitted to McDonnell Douglas for stress analysis approval. The repair was approved and performed, and the aircraft returned to service.

On January 5, 1979, Operational Occurrence Report No. 10-7901 was published by McDonnell Douglas. The publication contained descriptions of several DC-10 occurrences involving various aircraft systems, personnel injury, and the damage inflicted on the Continental Airlines DC-10. The report described the damage to the upper flange of the Continental aircraft and indicated that it occurred during maintenance procedures used at the time it was damaged. However, the way in which the damage was inflicted was not mentioned. The manufacturer had no authority to investigate air carrier maintenance practices and, therefore, accepted the carrier's evaluation of how the flange was damaged. Since the damage was inflicted during maintenance, 14 CFR 21.3 relieved McDonnell Douglas of any responsibility to report the mishap to the FAA. Although American Airlines was on the distribution list for Operational Occurrence Reports, testimony disclosed that the maintenance and engineering personnel responsible for the pylon maintenance were not aware of the report.

Continental Airlines discovered the damage to the second bulkhead in February 1979. Again the carrier evaluation indicated that the cause of the damage was related to personnel error, and that there was apparently no extensive effort to evaluate the engine-pylon assembly removal and reinstallation procedures. The bulkhead was also repaired using the procedure previously approved by McDonnell Douglas.

The carrier did not report the repairs that were made to the two bulkheads to return them to service, and there was no regulatory requirement to do so. What constitutes a major repair may be subject to interpretation, but what is to be reported is not. The bulkheads were not altered; they were repaired. Even had the repairs been classified by the carrier as major, 14 CFR 121.707(b) only requires that a report be prepared and kept available for inspection by a representative of the FAA. Second, the regulation does not indicate that the contents of the required report include a description of the manner in which the damage was inflicted. The regulation and the evidence indicated that the purpose of the reports was to permit the FAA to evaluate the end-product to insure that the basic design of the repaired or altered part had not been changed.

The Mechanical Reliability Reporting criteria of 14 CFR 121.703 requires the certificate holder to report "the occurrence or detection of each failure, malfunction, or defect concerning . . ." and then lists 16 criteria to which these apply. The FAA and apparently the aviation industry have traditionally interpreted 121.703 to apply to only service-related problems, which would therefore exclude reporting-of the flange damage caused by maintenance. In view of this interpretation, the board concludes that there is a serious deficiency in the reporting requirements which should be corrected.

Therefore, the safety board concludes that neither the air carrier nor the manufacturer interpreted the regulation to require further investigation of the damages or to report the damage to the FAA. However, the safety board views the omission of such requirements as a serious deficiency in the regulations.

McDonnell Douglas did not investigate Continental Airlines' maintenance procedures and accepted its finding that the damage was due to maintenance error. However, two months later McDonnell Douglas received the report that a second bulkhead was damaged, that the location and type of damage was almost identical to the damage inflicted on the first bulkhead, and that the damage was again due to maintenance error. McDonnell Douglas then had the opportunity to question whether maintenance error was the result of a procedural problem rather than accepting personnel error as the cause. They should have investigated the procedure and per-haps discovered the flaws within the procedure. However, they accepted the company's evaluation of cause and did not pursue the matter further.

The safety board, therefore, believes that the regulatory report-ing structure had and still has a serious deficiency. Damage to a component identified as "structurally significant" must be reported in a manner which will assure that the damage and the manner in which it is inflicted is evaluated, and the results of that evaluation disseminated to the operators and airframe manufacturers. Second, damage to a component of this type should be reported regardless of whether it was incurred during flight, ground operations, or main-tenance. Finally, damage suffered by these types of structures should be investigated by representatives of the operator, airframe manu-facturer, and the administrator.

*Surveillance of Industry Practices by
Federal Aviation Administration*

The Safety Board believes that the facts, conditions, and circum-stances of this accident and the information obtained during the investigation illustrate deficiencies in the aviation industry ranging from aircraft design through operations. The safety board recog-nizes that resource limitations prohibit the FAA from exercising rigid oversight of all facets of the industry. Therefore, the FAA must exercise its authority by insuring that aircraft designs do comply with regulations, that manufacturers quality control pro-grams are effective, that aircraft operators adhere to a proper maintenance program; and that operational procedures adopted by the carriers consider even unique emergencies which might be encountered.

In summary, the safety board recognizes that the overall safety record of the current generation of jet aircraft clearly indicates that the regulatory structure under which U.S. commercial aviation oper-ates and the industry's commitment to safety is basically sound. The safety board, however, is concerned that this accident may be indica-tive of a climate of complacency. Although the accident in Chicago on May 25 involved only one manufacturer and one carrier, the safety board is concerned that the nature of the identified deficiencies in design, manufacturing, quality control, maintenance and operations may reflect an environment which could involve the safe operation of other aircraft by other carriers.

Safety Recommendations

As a result of this accident, the National Transportation Safety Board has recommended that the Federal Aviation Administration:

- Issue immediately an emergency airworthiness directive to inspect all pylon attach points on all DC-10 aircraft by approved inspection methods. (Class I, Urgent Action) (A-79-41)
- Issue a telegraphic airworthiness directive to require an immediate inspection of all DC-10 aircraft in which an engine pylon assembly has been removed and reinstalled for damage to the wing-mounted pylon aft bulkhead, including its forward flange and the attaching spar web and fasteners. Require removal of any sealant which may hide a crack in the flange area and employ eddy-current or other approved techniques to ensure detection of such damage. (Class I, Urgent Action) (A-79-45)
- Issue a maintenance alert bulletin directing FAA maintenance inspectors to contact their assigned carriers and advise them to immediately discontinue the practice of lowering and raising the pylon with the engine still attached. Carriers should adhere to the procedure recommended by the Douglas Aircraft Company service bulletin which include removing the engine from the pylon before removing the pylon from the wing. (Class I, Urgent Action) (A-79-46)
- Issue a maintenance alert bulletin to U.S. certificated air carriers, and notify States that have regulatory responsibilities over foreign air carriers operating DC-10 aircraft, to require appropriate structural inspections of the engine pylons following engine failures involving significant imbalance conditions or severe side loads. (Class I, Urgent Action) (A-79-52)
- Incorporate in type certification procedures consideration of:
  (a) Factors which affect maintainability, such as accessibility for inspection, positive or redundant retention of connecting hardware and the clearances of interconnecting parts in the design of critical structural elements; and
  (b) Possible failure combinations which can result from primary structural damage in areas through which essential systems are routed. (Class II, Priority Action) (A-79-98)
- Insure that the design of transport category aircraft provides positive protection against asymmetry of lift devices during critical phases of flight; or, if certification is based upon demonstrated controllability of the aircraf t under condition of asymmetry, insure that asymmetric warning systems, stall warning systems, or other

critical systems needed to provide the pilot with information essential to safe flight are completely redundant. (Class II, Priority Action) (A-79-99)
- Initiate and continue strict and comprehensive surveillance efforts in the following areas:
  (a) Manufacturer's quality control programs to assure full compliance with approved manufacturing and process specifications; and
  (b) Manufacturer's service difficulty and service information collection and dissemination systems to assure that all reported service problems are properly analyzed and disseminated to users of the equipment, and that appropriate and timely corrective actions are effected. This program should include full review and specific FAA approval of service bulletins which may affect safety of flight. (Class II, Priority Action) (A-79-100)
- Assure that the maintenance review board fully considers the following elements when it approves an airline/manufacturer maintenance program:
  (a) Hazard analysis of maintenance procedures which involve removal, installation, or work in the vicinity of structurally significant components in order to identify and eliminate the risk of damage to those components;
  (b) Special inspections of structurally significant components following maintenance affecting these components; and
  (c) The appropriateness of permitting "On Condition" maintenance and, in particular, the validity of sampling inspection as it relates to the detection of damage which could result from undetected flaws or damage to structurally significant elements during manufacture or maintenance. (Class II, Priority Action) (A-79-101)
- Require that air carrier maintenance facilities and other designated repair stations:
  (a) Make a hazard analysis evaluation of proposed maintenance procedures which deviate from those in the manufacturer's manual and which involve removal, installation, or work in the vicinity of structurally significant components; and
  (b) Submit proposed procedures and analysis to the appropriate representative of the Administrator, FAA, for approval. (Class II, Priority Action) (A-79-102)
- Revise 14 CFR 121.707 to more clearly define "major" and "minor" repair categories to insure that the reporting requirement will

include any repair of damage to a component identified as "structurally significant." (Class II, Priority Action) (A-79-103)

- Expand the scope of surveillance of air carrier maintenance by:

  (a) Revising 14 CFR 121 to require that operators investigate and report to a representative of the administrator the circumstances of any incident wherein damage is inflicted upon a component identified as "structurally significant" regardless of the phase of flight, ground operation, or maintenance in which the incident occurred; and

  (b) Requiring that damage reports be evaluated by appropriate FAA personnel to determine whether the damage cause is indicative of an unsafe practice and assuring that proper actions are taken to disseminate relevant safety information to other operators and maintenance facilities. (Class II, Priority Action) (A-79-104)

16

The DC-10: A Special Report*

If there are in America or the world people who
hesitate to fly by DC-10, those people are, like
our airplane and our company, victims of a great
mass of misinformation and baseless speculation.
They have been misled, as we have been
maligned, by so much falsehood that the truth,
when it finally emerged, was scarcely recognized.
    —Sanford N. McDonnell, McDonnell Douglas
                    Annual Meeting, April 1980

TO SET THE RECORD STRAIGHT

There is no point, a rule as old as Aristotle tells us, in debating a question that can be settled simply by examining the facts.

It is ridiculous, in other words, to argue about whether the Yankees won the 1964 World Series. A glance at the record book will settle the matter immediately. Discussion will settle it never.

The McDonnell Douglas DC-10 became the subject of intense public debate on May 25, 1979, the date of a tragic crash at Chicago's O'Hare International Airport. Immediately after that crash, and for months thereafter, the DC-10 was often and prominently in the news.

---

* Reprinted, with editorial changes, from *The DC-10: A Special Report*, McDonnell Douglas Corporation, 1979.

At first, this interest in the DC-10 was appropriate. The Chicago accident was the worst in U.S. aviation history. The fact that it happened because an engine support pylon separated from the aircraft at the instant of takeoff gave rise to important—to urgent—questions:

What had caused the engine and pylon to separate from the airplane's wing?

Was there something wrong with the DC-10—some flaw, perhaps, in its design?

Was the DC-10 safe to fly?

The grounding of the DC-10 fleet shortly after the accident made these questions seem more important than ever. Even the validity of the DC-10's certification by the Federal Aviation Administration was questioned. People around the world suddenly wondered if the DC-10, the wide-cabin jetliner making more flights daily than any other, was in fact worthy of the public's trust.

Naturally, properly, discussion of the DC-10 continued as long as such questions remained unanswered. And not all of them were answered quickly. Some had to remain unsettled until complex and time-consuming studies had been completed. The last of them was not answered until December 1979, when the National Transportation Safety Board issued its final report on the Chicago accident, and January 1980, when the Federal Aviation Administration issued its final report on the DC-10.

The answers, when they emerged, were clear and conclusive. They proved that the DC-10 meets the toughest standards of aerospace technology.

They proved, too, that the Chicago accident did not result from any deficiencies of aircraft design, and that steps taken shortly after the accident had eliminated any possibility of recurrence.

And they proved that the grounding of the DC-10 fleet had been an unnecessary act based on incomplete information.

These are facts, not opinions. They have been established by teams of experienced, respected, independent technical experts using rigorous, objective methods. The availability of these facts has made further debate about the safety of the DC-10 a pointless exercise.

But good news often doesn't travel as far or as fast as bad. The vindication of an airplane, especially when it's based on thousands of pages of mathematically precise data, isn't as dramatic a story as a calamity. It lacks the human-interest appeal of a grounding that disrupts the whole world's air travel system.

And so answers sometimes never quite catch up with the questions that prompted them. Doubts linger—unnecessarily. Debate goes on—emptily.

The purpose of this booklet is to raise the basic, often-repeated questions about the DC-10 one final time. And, having raised them, to present the facts that answer them beyond the possibility of reasonable doubt.

And so to set the record straight.

THE BASIC QUESTIONS

Why did a DC-10's pylon and engine separate from the wing at Chicago?

The accident investigating team, including representatives of government agencies and independent outside experts, established that the pylon and engine separated from the airplane's wing because of a very large crack in what is called the horizontal flange of the pylon's aft bulkhead.

What was the origin of this crack?

The investigators found that the pylon could not have been damaged in this way while attached to the wing. It could not, in other words, have been damaged in flight. Rather, as the National Transportation Safety Board stated, it was "damaged by improper maintenance procedures." The maintenance procedures that caused the crack were of course banned by law as soon as the problem was identified.

Have changes in the pylon's structural design been ordered?

No. As the FAA stated, "there are no fundamental shortcomings in the design of the DC-10 wing pylon." In the six months after the Chicago accident, the DC-10 underwent a relatively new and highly sophisticated structural examination called damage tolerance analysis. This examination, in which the FAA, U.S. Air Force experts and outside consultants were involved, was more intensive than any previously used in the commercial aviation industry. It involved more than 5,000 pages of sophisticated calculations and established that the DC-10 wing pylon met not only the high standards in effect when the airplane was first certified but more recently developed standards as well. It confirmed the FAA's conclusion that "the DC-10 wing pylon is of sound design, material, specification, construction and performance." The FAA reported that the pylon is capable of at least twenty-five years of airline service. In May 1980 the FAA directed that two refinements be incorporated if, for any reason, a pylon should ever be removed again. These, it said, "should virtually eliminate the possibility" of damage resulting from improper removal and replacement of pylons. The first requires installation of two bolts with recessed rather than protruding heads to provide additional clearance at an attachment point. The second requires the use of a new maintenance tool which eliminates contact between the aft pylon bulkhead and the wing during pylon removal and installation. To further assure the greatest possible durability, the FAA also required that thrust links, which transfer the forward force of the engine to the wing, should be made of steel rather than titanium—a change introduced by McDonnell Douglas in the manufacture of DC-10s in 1975 despite evidence that titanium thrust links were strong enough for sixteen years of airline service.

Is the DC-10 pylon supported from the wing by a single quarter-inch bolt?

No. This misunderstanding arose when a broken bolt found shortly after the accident was shown on television in Chicago. Investigation soon confirmed that the bolt had failed as a result—not a cause—of the loss of engine and pylon. The DC-10 pylon is attached to the wing at four major attachment points each of which has dual connections, either of which possesses the strength to carry independently all of the forces which are transmitted through that point.

Why were DC-10s grounded?

Early in June 1979, inspections revealed pylon cracks in several DC-10s. No such cracks, however, had been detected when these same aircraft were inspected late in May almost immediately after the Chicago accident. Faced with this situation, and not waiting to determine the cause of the newly discovered cracks, on June 6 the FAA suspended the DC-10's type certificate—its license to fly in the

U.S. Further investigation, the FAA's January 1980 report acknowledges, established that "the first inspection reports had been erroneous in reporting the bulkheads to be free of cracks." Ironically, therefore, the grounding was the direct result not of pylon cracks but of a failure to detect existing cracks during the first round of postaccident inspections.

Are the DC-10's hydraulics systems effective and safe?

Yes. Questions about hydraulics were among the many raised in the immediate aftermath of the Chicago tragedy—and among the many that were soon settled. The DC-10 has three completely independent hydraulics systems each of which is powered by duplicate hydraulic pumps on one of the engines. These systems control the movable parts or control surfaces on the wings and tail, which direct the flight of the airplane. The plane remains flyable as long as any one of these systems continues to work properly. Separation of engine and pylon from the wing of the accident aircraft put one hydraulics system—the one powered by the lost engine—out of operation. A second system, though losing fluid because of damage resulting from pylon separation, was still operating when the crash occurred and would have remained in operation for several minutes. The third system was unaffected, and the airplane remained controllable. Hydraulics, in short, was not a contributing factor in the tragedy. The DC-10's hydraulics systems worked as they should have to provide redundant power even in the unlikely event of multiple damage. According to industry statistics they are among the most reliable hydraulics systems in use on transport aircraft today.

Is there a problem with the DC-10's wing slats?

No. Questions about slats, movable wing sections which extend to provide extra lift for takeoff and landing, were explored carefully by the National Transportation Board and the FAA as part of their investigations. Both agencies established that although the slats on the Chicago accident aircraft were asymmetrical after engine loss— meaning that the slats were in an extended position on one wing but retracted on the other—the airplane was flyable under such circumstances. The accident investigation was not the first time this question received attention. McDonnell Douglas had demonstrated the DC-10's ability to fly with asymmetrical slats almost a decade earlier, when the aircraft was being certified for airline service. It is the only wide-cabin airliner to have demonstrated this capability.

But weren't changes to the slats required after the accident?

Not to the slats themselves, and not to the mechanisms that control the movement of the slats. These were all found to be entirely satisfactory. The FAA did, however, order three changes in what is called the DC-10's stall warning system—the system by which the flight crew is informed if a wing is about to stall, or lose lift. These changes provide additional backup in the system which alerts the flight crew if a stalling situation is about to occur on either wing and an increase in aircraft speed is called for. The DC-10 stall warning system's "redundancy"—duplication to provide back-up security—exceeds industry standards for transport aircraft.

There were two other fatal DC-10 accidents in 1979 after the Chicago crash.

What caused them?

In the first, which occurred at Mexico City, the aircraft was flown down onto a runway that had been closed for repairs, where it collided with a truck. The other occurred during a special Antarctic flight; the aircraft hit a mountain while at very low altitude in a region where visibility was obscured and wind conditions were extreme and unpredictable. Investigating authorities have confirmed that no question of aircraft malfunction has arisen in connection with either case.

How does the DC-10's service record compare with those of other airliners?

All wide-cabin airliners, the DC-10 included, have excellent records. In numerous significant categories the DC-10's record is superior to the records of other, comparable planes. FAA records show that the DC-10 experiences engine shutdowns during flight far less often than the other two U.S.-built wide-cabin aircraft, and that the DC-10's service difficulty rate has been consistently lower for the past several years. In 1979 there were 1.8 service difficulty reports for every thousand DC-10 flights compared with 3.7 and 2.6 for the others. By all objective standards, the DC-10 has proven itself to be highly dependable.

Why, then, do we seem to hear so much about DC-10 problems?

It is demonstrably true that, since the Chicago tragedy and the intense media interest that it and the grounding stimulated, routine problems receive disproportionate coverage if and when they are associated with the DC-10 as opposed to other airliners. After the return of DC-10s to service in the summer of 1979, for example, there were numerous and prominent stories about DC-10 engine shutdowns—but virtually no attention to such shutdowns when they occurred on other planes. This was so despite the fact that, as the preceding answer indicates, the DC-10 has a superior record in the area of engine shutdowns when compared with other wide-cabin planes. In September 1979—when media attention to DC-10 shutdowns was at its peak—U.S. airlines reported twenty-one shutdowns on 747s and twelve on L-1011s. The comparable total for DC-10s was eight. But only the DC-10 shutdowns—less than a fifth of the total—made headlines. Another factor is an apparent lack of appreciation of the thoroughness with which the DC-10, after exhaustive and unprecedented study by leading experts in aviation technology, has been proved to be a well-designed, well-engineered and dependable aircraft. The sensational opinions of self-appointed, self-serving amateur "experts" have too often received more attention than the scientific conclusions of responsible public agencies and the accredited, professionally respected technical specialists brought into the official investigations. This has contributed to the truth being ignored and distorted.

MORE FACTS

The DC-10 is a carefully developed, meticulously engineered aircraft.

Forty-three months—more than three and a half years—separated McDonnell Douglas Corporation's decision to produce the DC-10 from certification of the aircraft by the FAA in 1971. This span of time is practically identical to that required for development of the 747, and of course it was backed up by McDonnell Douglas Corporation's decades of previous experience as a world-famous airplane builder.

Eighteen million engineering man-hours were invested by McDonnell Douglas in DC-10 development. This total included twelve million hours spent on design, four million on laboratory tests, and two million on flight tests. Subcontractors invested additional millions of hours.

In developing the DC-10, McDonnell Douglas conducted 14,000

hours of wind tunnel tests. Flight tests totaled 1550 hours—the equivalent of hundreds of flights across the U.S. Full-scale fatigue testing of the DC-10 provided the equivalent of 84,000 flights—forty years of airline service.

Since the DC-10 first entered airline service, McDonnell Douglas has invested an additional total of more than twenty-seven million engineering man-hours in product improvement. More than 5,000 additional hours of flight testing have been conducted.

More than twenty-five percent of all the members of the DC-10 design team had more than twenty years of experience in the transport aircraft industry.

It was as a result of such excellence that the DC-10 has been selected for use as an advanced tanker/cargo aircraft to increase the U.S. Air Force's aerial refueling and cargo airlift capabilities.

The DC-10 is a world leader in superior air transportation .

The DC-10 is used by more than forty airlines around the world. No other airliner of comparable size is used by so many.

The DC-10 carries more than a million passengers every week. No other wide-cabin airliner carries so many.

The DC-10 serves more than 160 cities around the world. No other wide-cabin airliner serves so many.

The DC-10 makes more than 750 flights daily. No other wide-cabin airliner makes more.

Since first entering service in 1971, the DC-10 has carried more than 275 million passengers.

The DC-10 continues a tradition of excellence.

McDonnell Douglas, the manufacturer of the DC-10, has been building airliners continuously for forty-six years—longer than any other company in the world.

McDonnell Douglas has delivered more than 15,500 transport aircraft—more than all other manufacturers combined.

One member of the DC family, the DC-3, is the-most-produced transport aircraft in history. Even today, more than four decades after the first DC-3 was introduced, many hundreds are still in service.

This heritage has helped make the DC-10 what the facts prove it to be: A great airplane with a demonstrated ability to provide great service to airline passengers everywhere.

AIR TRAVEL AT ITS BEST

The engineering excellence that makes the DC-10 outstandingly dependable has also helped to make it as attractive as any airplane ever built. Advanced aerospace technology has created opportunities for an entirely new level of passenger comfort. In developing the DC-10, McDonnell Douglas put all those opportunities to good use.

Travel by DC-10 is, for example, a remarkably quiet experience. The plane's three big engines, though much more powerful than those on earlier jetliners, also make much less noise. And the sound they do make is muffled by heavy cabin insulation.

The DC-10's passenger cabin windows are bigger than those on comparable jets. Double aisles in the eighteen-foot-wide cabin make it easy to move around, easy for attendants to provide the best in service.

Three separate air conditioning systems provide twenty cubic feet of fresh—not just recycled—air for every passenger every minute. All the air in the DC-10 cabin is replaced every three minutes.

Around the world,-DC-10s make about thirty flights an hour . . . one every two minutes . . . a great deal of comfortable, reliable service for vast numbers of people.

MARTIN CURD
LARRY MAY

17

Two Models of
Professional Responsibility*

This chapter presents two cases of putative negligence by professional engineers in an attempt to investigate more closely the scope and limits of responsibility in engineering practice. In these cases we invite you to judge what it is reasonable to expect of the engineers involved. We will appeal to your considered moral institutions and judgments while recognizing that you may not always agree with us. Morality, like legality, is subject to differences of opinion, especially in difficult cases. Underlying these disagreements, though, there is often a common core of shared judgments about fundamental issues and it is on these that we try to rely. The key issue in many of these cases is deciding what constitutes negligent

---

* Reprinted, with editorial changes, by permission from the Center for the Study of Ethics at IIT from *Professional Responsibility for Harmful Actions*, Copyright © 1984 by Illinois Institute of Technology, Kendall/Hunt Publishing Co., Dubuque, Iowa.

fault. In order to facilitate our analysis we propose the following simplified account of professional responsibility embodying a rather crude model of negligence.

THE MALPRACTICE MODEL OF PROFESSIONAL RESPONSIBILITY

A professional, S, is negligent and hence responsible for the harm he or she causes, if his or her behavior fits the following pattern:

1. As a member of his or her profession, S has a duty to conform to the standard operating procedures of his or her profession;
2. at time t, action X conforms to the standard operating procedures of S's profession;
3. S omits to perform X at time t;
4. harm is caused to some person, P, as a result of S's failure to do X—that is, if S had done X, then the harm to P would not have occurred.

Such a model of professional negligence leaves out several important factors. But before trying to improve this simple malpractice model we shall apply it to the first of two cases involving design defects in an aircraft. These two cases both concern crashes of DC-10s and are the main focus of our investigation of responsibility for harm in professional engineering.

There is convincing evidence that some of the engineers who designed and tested the DC-10 were negligent. For some of these engineers this negligence borders on recklessness. The McDonnell Douglas design team chose a cargo door latching device that was known to be less safe than the other latching mechanisms used in the aircraft industry for wide-bodied jets. Furthermore, the designers remained committed to the inferior latching system even after there was clear evidence that it posed a serious threat to human life. One of the chief components in establishing negligence is the foreseeability of the harm resulting from one's actions. The Paris crash of 1974 is one of the clearest cases on record of a major disaster that was completely foreseeable by the engineers and corporate managers involved. The authors of the documentary study, *Destination Disaster*, have concluded that "to some of its designers, the faults of the DC-10 had been obvious long before [1972]. And to judge by their written prophecies, neither Windsor nor the later tragedy outside Paris could have come as much of a surprise" (Eddy 1976, p. 165). Thus it is reasonable to expect the appropriate engineers and management of McDonnell Douglas to have anticipated the cargo door failure on the DC-10 and to have promptly modified this aspect of the aircraft's design. When one also considers that the standards in other parts of the aircraft industry dictated safer latching mechanisms, one sees why the primary responsibility for 346 deaths and the loss of the Turkish Airlines DC-10 can be attributed to the malpractice of the professionals employed by McDonnell Douglas.

As well as illustrating the malpractice model of negligence this case also reveals some of its limitations. It appears to have been standard operating procedure for the engineers involved in design and testing at McDonnell Douglas and Convair to defer to upper management. Even supervisory engineers such as Applegate did not voice their fears until after an accident had occurred. Given this practice it might appear that the individual engineers involved in the design and testing of the DC-10 had *not* acted negligently. This would be true, on the malpractice model outlined earlier, if the engineering profession does not recognize a professional duty to try to prevent management from proceeding with the manufacture of an unsafe aircraft or a duty to warn the public about the unsafe aircraft once it reaches the airlines. Some of the engineers involved in the DC-10 case argued that they had warned management about the problem and that it was then management's responsibility, not their own, if these warnings were ignored.

One obvious defect of the malpractice model suggested by this discussion and by tort cases in common law is that merely conforming to the standards of one's profession is not always a legitimate excuse for avoiding responsibility for harm. This leads us to propose a reasonable care model of negligence which falls somewhere between the malpractice model and the doctrine of strict liability. As indicated earlier, strict liability eliminates the fault criterion entirely and stipulates that anyone who causes or causally contributes to a harm is responsible for that harm regardless of foreseeability, reasonableness or fault. The middle ground we propose is not a standard of strict liability but it recognizes a standard that may be higher than that required by one's profession. This model superimposes a standard of reasonableness as seen by a normal, prudent nonprofessional over that of the reasonable professional. Thus where the nonprofessional would exercise more care than the average professional, then the extra care is required of the professional. But, if the average person would not act as carefully as the professional (perhaps due to ignorance), then the standard of due care is determined by the behavior of the average professional. Thus, we would amend the earlier malpractice model as follows.

THE REASONABLE CARE MODEL OF
PROFESSIONAL RESPONSIBILITY

A person, S, is responsible for the harm he or she causes when his or her conduct fits the following pattern:

1. As a member of a profession, S has a duty to conform to the standard operating procedures of his or her profession, unless those standards are lower than those that a nonprofessional would adopt in a given situation, in which case S has a duty to conform to the higher standard;
2. at time t, action X conforms to the standard of reasonable care defined in (1);
3. S omits to perform X at time t;
4. harm is caused to some person, P, as a result of S's failure to do X.

The following case illustrates the importance of this amended model of professional responsibility.

Unlike the previous case of the Turkish Airlines crash, it is not at all clear that the design engineers at McDonnell Douglas were responsible for the disaster in Chicago.

Though the Safety Board criticized several design features of the DC-10 (principally the absence of a slat-locking mechanism and the aircraft's lower degree of fail-safe back-up systems as compared with the 747 and the TriStar), the primary responsibility was said to rest with the improper maintenance procedures of the airlines that owned and operated the plane which crashed. Admittedly, McDonnell Douglas was found guilty of lapses in quality control that led to defective pylons in the planes purchased by United Airlines. But these particular defects, while serious in themselves, were not causally relevant to the crash of the American Airlines plane in Chicago. Judged by the standards widely accepted in the aircraft industry prior to the 747 and the TriStar, the DC-10 was an adequate and safe design (leaving aside the cargo door locking mechanism). Judged by the standards *created* by the 747 and the TriStar, the DC-10 was inferior with regard to safety in extreme and improbable situations. But with the larger numbers of passengers carried by a single flight of a wide-bodied jet, even relatively small margins of risk become unacceptable.

Both the 747 and the TriStar possess safety features lacking in the DC-10 which, had the engineers at McDonnell Douglas incorporated them into their own design, might have saved Flight 191 from disaster. Some commentators have suggested that the inferior design of the DC-10 from the point of view of safety resulted from its overly hasty construction and that this was a clear case of corporate irresponsibility. We will assess this claim and, for us, the more interesting contention that in Case II as in Case I, the design engineers were responsible for the harms caused by their (supposed) professional negligence.

The main difference between Cases I and II is that, in the first but not in the second, the engineers and management involved clearly violated the norms of professional conduct in the aircraft manufacturing industry when they deliberately went ahead with the cargo door latching mechanism that they should have known (and almost certainly did know) was potentially dangerous. But in Case II, there were no such striking violations of professional standards. The FAA, for example, did not consider the DC-10 control systems to be significantly less safe than those of the 747 and the TriStar. But is this sufficient to relieve the designers of the DC-10 of all responsibility for the Chicago crash? The judgment depends on what is considered reasonable risk in the aircraft industry. The FAA, like most governmental regulatory agencies, sets only *minimal* standards of safety. We believe that reasonable risk is not the same as minimally acceptable risk, especially in industries where there is a great potential for harm. Of crucial relevance to our assessment is that McDonnell Douglas had had one of its DC-10s crash near Paris five years before. We hold that when a hazardous design defect has been found in one of the products of a company, that company then has a duty to make sure that the product is safe, *not just in respect of the known defect*, before the product is allowed back into the public sector. In one way at least, Cases I and II are quite similar: Boeing and Lockheed had opted for safer designs than had McDonnell Douglas. After the Paris crash, it was no longer possible for McDonnell Douglas to say that while the design of the DC-10 was different from that of its competitors, it was nevertheless within the limits of *reasonable* design risk.

A second difference between Cases I and II is that in the second, but not in the first, the negligence of the airline maintenance crews was a contributing cause of the accident. As mentioned earlier, in legal theory the negligent acts of another which are subsequent to the original acts relieve the first actor of responsibility, if the harm would not have occurred without the second intervention and *if those subsequent acts of negligence could not reasonably have been anticipated by the first actor.*

Was it reasonably foreseeable that the airline maintenance crews would omit to disassemble the pylon from the engine while servicing it? We believe it was.

It has been estimated that it would have taken an extra 200 man-hours per engine to service them in the manner recommended by McDonnell Douglas. More importantly, McDonnell Douglas knew that maintenance crews were using the less time-consuming and more hazardous procedure. Continental Airlines and American Airlines were both using forklift trucks to remove the entire engine and pylon assembly as a unit. In December, 1987, and again in February, 1979 (several months prior to the Chicago crash), cracked flanges were discovered in planes belonging to Continental that had been serviced in this way. The Safety Board concluded: "Neither the air carrier (Continental) nor the manufacturer (McDonnell Douglas) interpreted the (FAA) regulation to require that it further investigate or report the damage to the FAA" (Newhouse 1982, p. 87). Once the McDonnell Douglas engineers saw that the risk of cracks in the flange was no longer so remote, they should have taken action to reduce the chances of a serious accident. While it is undeniable that the members of the maintenance staff acted wrongly and were thus responsible for the cracks in the pylons, McDonnell Douglas must assume part of the blame as well.

A third difference between the two cases is that in the first, but not in the second, there is substantial evidence that individual engineers within McDonnell Douglas had recognized the safety problems involved (the cargo door latching mechanism; the inadequate provision for mitigating the effects of sudden decompression) and were convinced that they constituted an unacceptable risk. As a result these engineers must share some of the responsibility for the ensuing harm since they clearly foresaw what might happen and they were in a position to try to prevent these hazards from endangering the public. The model of individual responsibility within a corporation presupposed by this judgment is defended below. Suffice it to say that only on the second model of negligence, the reasonable care model, is it true that the engineers involved could be held responsible for the Paris crash because they failed to prevent the DC-10 from reaching the market in its unsafe condition.

The reasonable care model imposes a stronger duty of care on professional engineers than does the malpractice model. In Case II the design team's initial decision not to make the flange stronger was reasonable given the extremely remote chance of harm resulting. But after the cracks in the flange were reported, and after it became clear that the reduced maintenance crews at American and Continental were not maintaining the engines and pylons properly, then, on the reasonable care model, it became the duty of individual engineers who knew of this to try to prevent the harm that was now far more likely to occur. This is one of the consequences of superimposing the reasonable care standard on the malpractice model.

One advantage of the reasonable care model is that it discourages engineers from believing that their professional responsibility is diminished by upper management decisions within their corporations. Aside from working to preserve professional integrity, this places the responsibility for eliminating or reducing hazards on those best able to anticipate them. Many upper management personnel are not engineers and even when they are, competing pressures, such as the desire to increase profits or protect the company, may compromise their professional judgment. For them reasonable risk will always be defined, at least partially, in terms of increased cost to the company. While the pursuit of profit is a legitimate motive in any business, it should not play so large a role when potentially disastrous consequences might result from cost cutting. In *MacPherson v. Buick Motor Company* (1916), the New York Court of Appeals recognized that products were too complicated to be understood by the average consumer. Thus, it was argued, the manufacturer has a duty to insure that its products are manufactured safely. For the same reasons, we contend that professional engineers should get actively involved when manufacturing companies put profits ahead of concern for public safety. The engineer is most likely to know the hazards that the unsuspecting public will encounter with products such as aircraft. Thus the engineer is the one who should take responsibility for their careful design.

THE 1989 SIOUX CITY CRASH

18

National Transportation Safety Board Report on the 1989 Sioux City Crash*

EXECUTIVE SUMMARY

On July 19, 1989, at 1516, a DC-10-10, N1819U, operated by United Airlines (UAL) as flight 232, experienced a catastrophic failure of the number two tail-mounted engine during cruise flight [see fig. 19.1]. The separation, fragmentation and forceful discharge of stage one fan rotor assembly parts from the number two engine led to the loss of the three hydraulic systems that powered the airplane's flight controls [see figs. 19.2, 19.3, and 19.4]. The flightcrew experienced severe difficulties controlling the airplane, which subsequently crashed during an attempted landing at Sioux Gateway Airport, Iowa. There were 285 passengers and eleven crewmembers onboard. One flight attendant and 110 passengers were fatally injured.

---

* Reprinted, with editorial changes, from *National Transportation Safety Board Report, NTSB-AAR-90-06,* (1990).

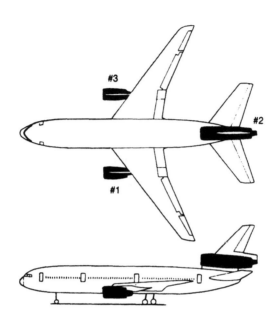

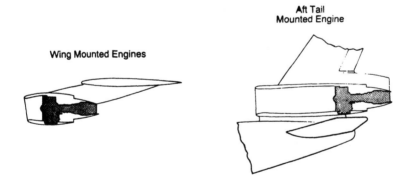

FIG. 18.1
DC-10 airplane view illustrated with engine arrangement

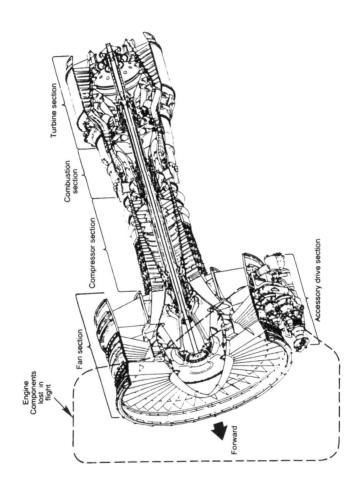

FIG. 18.2
CF6-6 engine cutaway

250

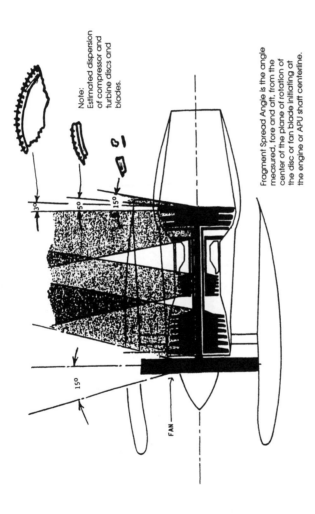

FIG. 18.3
Estimated Path of Fragments

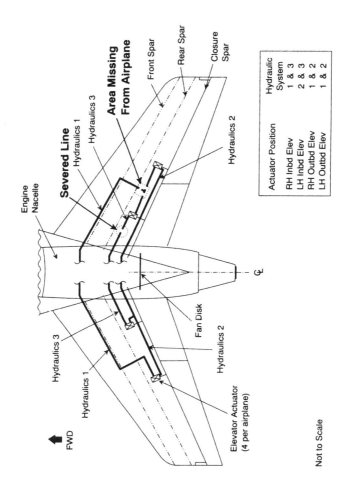

Actuator Position	Hydraulic System
RH Inbd Elev	1 & 3
LH Inbd Elev	2 & 3
RH Outbd Elev	1 & 2
LH Outbd Elev	1 & 2

FIG. 18.4

N1819U, planform of horizontal stabilizer hydraulic system damage

The National Transportation Safety Board determines that the probable cause of this accident was the inadequate consideration given to human factors limitations in the inspection and quality control procedures used by United Airlines' engine overhaul facility which resulted in the failure to detect a fatigue crack originating from a previously undetected metallurgical defect located in a critical area of the stage 1 fan disk that was manufactured by General Electric Aircraft Engines. The subsequent catastrophic disintegration of the disk resulted in the liberation of debris in a pattern of distribution and with energy levels that exceeded the level of protection provided by design features of the hydraulic systems that operate the DC-10's flight controls.

Jim Burnett, Member, filed the following dissenting statement on the probable cause:

> I believe that the probable cause of the accident was:
> (1) the manufacture by General Electric Aircraft Engines (GEAE) of a metallurgically defective titanium alloy first stage fan disk mounted on the aircraft's number two engine and the failure to detect or correct the condition;
> (2) the failure of United Airlines to detect a fatigue crack which developed from the defect and ultimately led to a rupture of the disk and fragmentation damage that disabled the airplane's hydraulically powered fight control systems; and
> (3) the failure of the Douglas Aircraft Company's (Douglas) design of the airframe to account for the possibility of a random release and dispersion of engine fragments following a catastrophic failure of the number two engine.

Flight Simulator Studies

As a result of the accident, the safety board directed a simulator reenactment of the events leading to the crash. The purpose of this effort was to replicate the accident airplane dynamics to determine if DC-10 flight crews could be taught to control the airplane and land safely with no hydraulic power available to actuate the flight controls. The simulator exercise was based only on the situation that existed in the Sioux City accident—the failure of the number two (center) engine and the loss of fluid for all three hydraulic systems.

The DC-10 simulator used in the study was programmed with the aerodynamic characteristics of the accident airplane that were validated by comparison with the actual flight recorder data. DC-10 rated pilots, consisting of line captains, training clerk airmen, and production test pilots were then asked to fly the accident airplane profile. Their comments, observations, and performance were recorded and analyzed. The only means of control for the flight crew was from the operating wing engines. The application of asymmetric power to the wing engines changed the roll attitude, hence the heading. Increasing and decreasing power had a limited effect on the pitch attitude. The airplane tended to oscillate about the center of gravity (CG) in the pitch axis. It was not possible to control the pitch oscillations with any measure of precision. Moreover, because airspeed is primarily determined by pitch trim configuration, there was no direct control of airspeed. Consequently, landing at a predetermined point and airspeed on a runway was a highly random event.

Overall, the results of this study showed that such a maneuver involved many unknown variables and was not trainable, and the degree of controllability during the approach and landing rendered a simulator training exercise virtually impossible. However, the results of these simulator studies did provide some advice that may be helpful to flightcrews in the extremely unlikely event they are faced with a similar situation. This information has been presented to the industry by the Douglas Aircraft Company in the form of an "All DC-10 Operators Letter." In addition to discussing flight control with total hydraulic failure, the letter describes a hydraulic system enhancement mandated by an FAA airworthiness directive.

Performance of UAL 232 Flightcrew

Because of the loss of the three hydraulic systems, the flightcrew was confronted with a unique situation that left them with very limited control of the airplane. The only means available to fly the airplane was through manipulation of thrust available from the number one and number three engines. The primary task confronting the flightcrew was controlling the airplane on its flightpath during the long period (about sixty seconds) of the "phugoid" or pitch oscillation. This task was extremely difficult to accomplish because of the additional need to use the number one and number three power levers asymmetrically to maintain lateral (roll) control coupled with the need to use increases and decreases in thrust to maintain pitch control. The flightcrew found that despite their best efforts, the airplane would not maintain a stabilized flight condition.

Douglas Aircraft Company, the FAA, and UAL considered the

total loss of hydraulic-powered flight controls so remote as to negate any requirement for an appropriate procedure to counter such a situation. The most comparable maneuver that the flightcrew was required to accomplish satisfactorily in a DC-10 simulator was the procedure for managing the failure of two of the three hydraulic systems; however, during this training, the remaining system was available for movement of the flight controls.

The CVR recorded the flightcrew's discussion of procedures, possible solutions, and courses of action in dealing with the loss of hydraulic system flight controls, as well as the methods of attempting an emergency landing. The captain's acceptance of the check airman to assist in the cockpit was positive and appropriate. The safety board views the interaction of the pilots, including the check airman, during the emergency as indicative of the value of cockpit resource management training, which has been in existence at UAL for a decade.

The loss of the normal manner of flight control, combined with an airframe vibration and the visual assessment of the damage by crewmembers, led the flightcrew to conclude that the structural integrity of the airplane was in jeopardy and that it was necessary to expedite an emergency landing. Interaction between the flightcrew and the UAL system aircraft maintenance network (SAM) did not lead to beneficial guidance. UAL flight operations attempted to ask the flightcrew to consider diverting to Lincoln, Nebraska. However, the information was sent through flight dispatch and did not reach the flightcrew in time to have altered their decision to land at the Sioux Gateway Airport.

The simulator reenactment of the events leading to the crash landing revealed that line flightcrews could not be taught to control the airplane and land safely without hydraulic power available to operate the flight controls. The results of the simulator experiments showed that a landing attempt under these conditions involves many variables that affect the extent of controllability during the approach and landing. In general, the simulator reenactments indicated that landing parameters, such as speed, touchdown point, direction, attitude, or vertical velocity could be controlled separately, but it was virtually impossible to control all parameters simultaneously.

After carefully observing the performance of a control group of DC-10-qualified pilots in the simulator, it became apparent that training for an attempted landing, comparable to that experienced by UA 232, would not help the crew in successfully handling this problem. Therefore, the Safety Board concludes that the damaged DC-10 airplane, although flyable, could not have been successfully landed on a runway with the loss of all hydraulic flight controls. The Safety Board believes that under the circumstances the UAL flightcrew performance was highly commendable and greatly exceeded reasonable expectations.

Airplane Flight Controls and Hydraulics—Description

Primary flight controls on the DC-10-10 consist of inboard and outboard ailerons, two-section elevators, and a two-section rudder. Secondary flight controls consist of leading edge slats, spoilers, inboard and outboard flaps, and a dual-rate movable horizontal stabilizer. Flight control surfaces are segmented to achieve redundancy. Each primary and secondary control surface is powered by two of three independent hydraulic systems.

The number one hydraulic system provides power to the right inboard aileron and the left outboard aileron, the right inboard and outboard elevators, the left outboard elevator, the upper rudder, the horizontal stabilizer trim, and the captain's brake system. The number two hydraulic system provides power to the right outboard aileron and the left inboard aileron, the inboard and outboard elevators on the left side, the outboard elevator on the right side, and the lower rudder. It also provides power to the isolated closed-loop system that operates the upper rudder. The number three hydraulic system provides power to the right inboard and outboard aileron and the left inboard aileron, the inboard elevators on the right and left side, horizontal stabilizer trim, and the first officer's brake system. It also drives an isolated closed-loop system that powers the lower rudder actuator. These closed-loop arrangements allow for operation of the remaining parts of hydraulic systems number two and number three in the event of damage to the rudder hydraulic system.

The three independent, continuously operating hydraulic systems are intended to provide power for full operation and control of the airplane in the event that one or two of the hydraulic systems are rendered inoperative. System integrity of at least one hydraulic system is required—fluid present and the ability to hold pressure—for continued flight and landing; there are no provisions for reverting to manual flight control inputs.

Each hydraulic system derives its power from a separate engine, with a primary and a reserve engine-driven pump providing hydraulic pressure. Either of these pumps can supply full power to its system. Backup power is provided by two reversible motor pumps, which transmit power from one system to another without fluid interconnection. This backup power system activates automatically without requiring flightcrew control, if fluid is still available in the unpowered system.

Electrical power can be used to drive either of two auxiliary pumps provided for the number three hydraulic system. In an emergency situation where the engine-driven pumps are inoperative, an air-driven generator can be deployed into the airstream to supply electrical power to one of these auxiliary pumps.

The hydraulic components and piping are physically separated to minimize the vulnerability of the airplane to multiple hydraulic system failures in the event of structural damage. The number one hydraulic system lines run along the left side of the fuselage to the rear of the airplane and along the front spar of the horizontal stabilizer and the vertical stabilizer. The number two hydraulic system lines are routed from the center engine along the rear spar of the horizontal and vertical stabilizers. The number three hydraulic system lines run along the right side of the fuselage to the tail area and along the rear spar of the horizontal stabilizer. The number two hydraulic system lines are not routed forward of the rear wing spar, in order to isolate them from wing engine fragmentation, and number three hydraulic system lines in the tail section are not routed aft of the inboard elevator actuators in order to minimize exposure to possible engine fragmentation damage from the tail-mounted engine.

The DC-10-10 hydraulic system was designed by the manufacturer and demonstrated to-the FAA to comply with 14 CFR 25.901, which in part specified that, "no single [powerplant] failure or malfunction or probable combination of failures will jeopardize the safe operation of the airplane. . . ."

Certification Requirements—Aircraft

Certification requirements for the DC-10-10 were specified in the 14 CFR; Part 25 Airworthiness Standards: Transport Category Airplanes dated February 1, 1965, with Amendments one through twenty-two and Special Condition 25-18-WE-7, dated January 7, 1970. Part twenty-five, paragraph 25.903(d) governed turbine powerplant installations. This paragraph stated that:

Unless the engine type certification specifies that the engine rotor cases can contain damage resulting from rotor blade failure, turbine engine powerplant installations must have a protection means so that rotor blade failure in any engine will not affect the operation of remaining engines or jeopardize continued safety. In addition, design precautions must be taken to minimize the probability of jeopardizing safety if an engine turbine rotor fails unless:

(1) The engine type certificate specifies that the turbine rotor can withstand damage-inducing factors (such as those that might result from abnormal rotor speed, temperature or vibration); and (2) The powerplant systems associated with engine devices, systems and instrumentation give reasonable assurance that those engine operating limitations that adversely affect turbine rotor structural integrity will not be exceeded.

Special Condition 2S-18-WE-7 stated that, "In lieu of the requirement of (paragraph) 25.903(d)(1), the airplane must incorporate design features to minimize hazardous damage to the airplane in the event of an engine rotor failure or of a fire which burns through the engine case as a result of an internal engine failure."

Special Condition 25-18-WE-7 was imposed by the FAA as part of certification of the DC-10-10 because FAR 25.903(d) was in the process of being revised and the applicable airworthiness requirement did not contain adequate or appropriate safety standards for the DC-10. In response to the special condition requirements, on July 1, 1970, Douglas responded by supplying information to the FAA that indicated the powerplants and associated systems were isolated and arranged so that the probability of the failure of any one engine or system adversely affecting the operations of the other engines or systems was "extremely remote." The response also noted that hydraulic system design considerations demonstrated compliance with the special conditions. The FAA responded on July 17, 1970, that the review of Douglas' compliance was complete and that the requirements of the applicable regulations and special conditions were satisfied. Amendment twenty-three was adopted after DC-10 certification and included the revised FAR 25.903(d). FAR 25.903(d)(1) mandated "incorporation of design features to minimize the hazards to the airplane in the event of a rotor (disk) failure."

FAA Order No. 8110.11 dated November 19, 1975, entitled "Design Considerations for Minimizing Damage Caused by Uncontained Aircraft Turbine Engine Rotor Failures" was distributed internally to various FAA offices.

Specific FAA-prepared advisory methods for compliance with 25.903(d) were not published until March 3, 1988, following a safety board recommendation on uncontained rotor separation events. Advisory circular (AC) 20-128 entitled "Design Considerations for Minimizing Hazards Caused by Uncontained Turbine Engine and Auxiliary Power Unit Rotor and Fan Blade Failure" set forth suggested methods for compliance with the FAR. In this AC, the FAA defines potential fragment spread angles that should be considered in the design of the aircraft to minimize the hazards associated with uncontained rotor failures. Predicted piece size and energy levels are discussed. Further, this AC proposed that critical components, such as critical control systems and hydraulic systems, be located outside the area of debris impact, as determined by the spread angle and fragment energy levels. If this is not possible, shields or deflectors should be considered to minimize the hazard of the uncontained debris.

FAA Order 8110 11 contained much of the same information as contained in AC 20-128, including fragment spread angles and the suggested use of shields or deflectors. Neither FAA Order 8110.11 nor AC 20-128 were effective at the time of certification of the DC-10-10.

Hydraulic System Enhancement

On September 15, 1989, Douglas Aircraft Company announced development of design enhancements to the DC-10's hydraulic system that would preserve adequate flight control if a catastrophic in-flight event in the empennage [tail] of the airplane damages all three hydraulic systems. The enhancements consist of three separate installations: (1) an electrically operated shutoff valve in the supply line and a check valve in the return line of the number three hydraulic system, (2) a sensor switch in the number three hydraulic reservoir, and (3) an annunciator light in the cockpit to alert the crew to the activation of the shutoff valve.

The shutoff valve is located in the empennage forward of the horizontal stabilizer. Normally open, the valve will close automatically if the sensor switch detects hydraulic fluid dropping below a preset level in the number three reservoir. The switch will also illuminate the alert light in the cockpit. If severe damage results in a breach of the number three hydraulic system anywhere in the aircraft, the shutoff valve will stop fluid flow through the number three hydraulic system routed through the tail. The hydraulic system enhancement is intended to provide the crew with longitudinal control by stabilizer trim input at one-half rate and lateral control through right inboard, right outboard, and left inboard aileron deflection, and with slats (but no flaps) in the event that an aircraft sustains damage similar to the damage sustained by flight 232. In addition, fluid for operation of the spoiler panels, brakes, nose wheel steering, landing gear, and lower rudder is preserved. The enhancement was mandated by FAA AD 90-13-07 effective July 20, 1990. The AD requires incorporation of the hydraulic system enhancement in all DC-10 airplanes on or before July 20, 1991.

In addition to the previously discussed shutoff valve system, Douglas also offered a system that incorporated flow-limiting fuses in the number three hydraulic system. Service bulletins were issued by Douglas to cover the installation of either system. AD 90-13-07 required that CF6-6-equipped DC-10 airplanes (DC-10-10 and DC-10-10F) have either the shutoff valve or flow-limiting fuses installed within six months of the AD issue date. All other models of the DC-10 were required to be modified with the shutoff valve within twelve months. The AD also required that if flow-limiting fuses were installed, the airplane must also have the shutoff valve installed within twelve months. The operators had the option of leaving the fuses in the system if they had been installed.

Douglas has incorporated the enhanced hydraulic system in the MD-11. All MD-11 airplanes will be manufactured with the shutoff valve system installed.

*Hydraulic Systems/Flight Control Design
Concept and Certification*

The three hydraulic systems installed on the DC-10 are physically separated in a manner that is intended to protect the integrity of the systems in a single-event-failure. Hydraulic fluid is isolated between the three independent systems and alternate motive systems and auxiliary systems are provided.

During the investigation of this accident, the safety board reviewed alternative flight control system design concepts for wide-body airplanes. The concept of three independent hydraulic systems, as installed on the DC-10, is not unique. Boeing and Airbus have three such systems on some of their most recently certified models. Lockheed and Boeing have also provided four independent systems on some of their wide-body airplanes. The safety board can find no inherent safety advantage to the installation of additional independent hydraulic systems for flight controls beyond those currently operating in today's fleet. However, the safety board believes that backup systems to the primary hydraulic systems should be developed and included in the initial design for certification. Such backup systems are particularly important for the coming generation of wide-body airplanes. Manual reversion flight control systems are quite likely impractical because of the power requirements to deflect large control surfaces that are heavily loaded. Therefore, the safety board recommends that the FAA encourage continued research and development into backup flight control systems that employ an alternative source of motive power.

Additional design precautions could have been taken by Douglas if the potential effect of the distribution pattern and fragment energy levels had been predicted. Engine manufacturers should provide such data to the airframe manufactures who can then incorporate measures to counter the effects into the airframe design. The problem is complicated by many factors, including the interaction of the nacelle design, engine pylon design, and supporting airframe structure.

During the UAL 232 accident sequence, once the fan disk failed and the pieces began to escape the confines of the containment ring, the dispersion of rotor disk and fan blade fragments was altered by contact with both engine components and the airplane structure. The safety board did not attempt to determine the specific origin or trajectory of each fragment that damaged the airplane in flight. For accident prevention purposes and in the course of making safety recommendations, it was sufficient to recognize that catastrophic damage from the failure of rotating parts can originate from any fragment source with sufficient energy to penetrate the airplane's structure.

The safety board considers in retrospect that the potential for hydraulic system damage as a result of the effect of random engine debris should have been given more consideration in the original design and certification requirements of the DC-10 and that Douglas should have better protected the critical hydraulic system(s) from such potential effects. As a result of lessons learned from this accident, the hydraulic system enhancement mandated by AD 90-13-07 should serve to preclude loss of flight control as a result of a number two engine failure. Nonetheless, the Safety Board is concerned that other aircraft may have been given similar insufficient consideration in the design for redundancy of the motive power source for flight control systems or for protecting the electronic flight and engine controls of new generation aircraft. Therefore, the safety board recommends that the FAA conduct system safety reviews of currently certificated aircraft in light of the lessons learned in this accident to give all possible consideration to the redundancy and protection of power sources for flight and engine controls.

Stage One Fan Disk Historical Data

The stage one fan disk, part number (P/N) 9137M52P36,[1] S/N MPO 00385, was processed in the manufacturing cycle at the GEAE-Evendale, Ohio, factory from September 3 to December 11, 1971. It was installed as a new part in engine S/N 451-251 in the GEAE production assembly facility in Evendale. The engine was shipped to Douglas Aircraft Company on January 22, 1972, where it was installed on a new DC-10-10.

During the next seventeen years, the engines in which this stage one fan disk were installed were routinely overhauled and the fan module was disassembled. The disk was removed on the following dates for inspection: September 1972, November 1973, January 1976, June 1978, February 1982 and February 1988. This disk was accepted after each of six fluorescent penetrant inspections (FPI).[2] Five of the six inspections were performed at the UAL CF6 Overhaul Shop in San Francisco, California. One of them was performed at the GEAE Airline Service Department in Ontario, California, in 1973. At the time of the accident, the stage one fan disk had accumulated 41,009 hours and 15,503 cycles since new. The last shop visit in February 1988, was 760 flight cycles before the accident, and FPI was performed at that time. The engine had been removed because of corrosion-in the high pressure turbine (HPT) stage one nozzle guide vanes. At that time, the stage one fan disk had accumulated 38,839 hours and 14,743 cycles since new. Following this inspection, the disk was installed in engine S/N 451-243, the number two engine on the accident airplane.

Operator Inspection Program and Methods

Maintenance records indicated that the stage one fan disk, the fan booster disk, the fan shaft, and the number one bearing had been inspected in accordance with the UAL maintenance program and the GEAE CF6-6 shop manual. The records search also showed that none of the engines in which the fan disk had been installed had experienced an overspeed or bird strike. There were no items in the prior three months' flight records relating to the fan components.

The stage one fan disk records indicated that the disk had been through six detailed part inspections in its lifetime, each of which included FPI of the entire disk. All of them had been stamped and accepted by the inspectors with no crack indications observed. The last inspection was about one year prior to the accident. All the records examined, as well as the life history and tracking methods, appeared to be in accordance with the FAA-approved UAL maintenance program.

Based on the evaluations and contributions from GEAE, UAL, and FAA, the Safety Board believes that the GEAE predictions of crack size more closely represent actual conditions. That is, GEAE fracture mechanics predictions indicate that, at the time of the last inspection, the length of the crack was almost one-half inch along the bore surface.

The portion of the fatigue crack around the origin that was discolored was slightly less than one-half-inch long along the bore surface. This size corresponds reasonably well to the size of the crack predicted by the GEAE fracture mechanics evaluation. Therefore, the Safety Board concludes that the discolored area marks the size of the crack at the time of the last inspection and that processing steps during the inspection created the discoloration.

During FPI inspection, a crack the size of the discolored region should have a high probability of detection, presuming that a proper inspection was conducted. At the time of the inspections prior to the most recent inspection in April 1988, the crack in the disk would have been much smaller. However, the GEAE fracture mechanics evaluation indicated that the surface length of the crack during several of the inspections prior to April 1988 was such that the crack would normally have been detectable by FPI. The Safety Board recognizes, however, that the unique metallurgical properties of the origin area may have altered the detectability of the crack during these inspections.

Analytical procedures performed on the fracture face of the segment of the rotor disk and water washings from this surface showed the presence of di and triphenyl phosphates, compounds present in FPI fluid similar to that used to inspect the disk prior to the failure. This unique combination of chemicals shows that the crack existed at the time of this inspection and that the crack was sufficiently open so that the FPI fluid entered the crack. Based on this finding and the conclusion from metallurgical analysis that the crack was approximately one-half-inch long on the surface of the bore of the rotor disk at the time of last inspection, the Safety Board concludes that the crack was detectable at the time of last inspection with FPI fluid. However, the crack was not detected and consequently the rotor disk was considered to be free of flaws and was accepted as a serviceable part.

A review of the inspection process suggests several explanations for the inspector's failure to detect the crack. It is possible that the inspector did not adequately prepare the part for inspection or that he did not rotate the disk, as it was suspended by a cable, to enable both proper preparation and subsequent viewing of all portions of the disk bore, particularly the area hidden by the suspension cable/hose. It is also possible that loose developer powder, which could have dropped from the suspension cable, obscured the crack sufficiently to prevent its recognition as a flaw. Finally, inspection experience indicates that certain areas of CF-6 disks, because of their geometry, frequently show large FPI indications and that other areas rarely do so. One such area of frequent indications is around the perimeter of the disk near the dovetail posts. By contrast, the central bore area apparently has rarely produced FPI indications. Thus, it is possible that the inspector did not consider the bore area a critical area for inspection, as stated in UAL's inspection directives, and that he gave the bore area only cursory attention, thereby reducing the likelihood that a crack would be detected. Any of these possibilities, or some combination of them, could have contributed to nondetection of the crack in this case.

The UAL maintenance program is comprehensive and based on industry standards. The company's inspection requirements for the CF6-6 stage one fan disk are generally consistent with other airline practices and comply with federal regulations. Further, UAL's procedures for selecting, training, and qualifying NDI personnel are also consistent with industry practices. However, it is clear that the adequacy of the inspections is dependent upon the performance of the inspector. That is, there are human factors associated with NDI processes that can significantly degrade inspector performance. Specifically, NDI inspectors generally work independently and receive very little supervision. Moreover, there is minimum redundancy built into the aviation industry's FPI process to prevent human error or other task or workplace factors that can adversely affect inspector performance. Because of these and other similar factors, the Safety Board is concerned that NDI inspections in general, and FPI in particular, may not be given the detailed attention that such a critical process warrants.

NOTES

1. Original P/N 9010M27P10 was superseded when the disk was modified during a GEAE shop visit in 1973. The fan blade dovetail slots were rebroached at that time.

2. Fluorescent penetrant inspection (FPI) is the accepted industry inspection technique for interrogating nonferrous (nonmagnetic) component surfaces for discontinuities or cracks. The technique relies on the ability of a penetrant (a low-viscosity penetrating oil containing fluorescent dyes) to penetrate by capillary action into surface discontinuities of the component being inspected. The penetrant fluid is applied to the surfce and allowed to penetrate into any surface discontinuities. Excess penetrant is then removed from the component surface. A developer is then applied to the component surface to act as a blotter and draw the penetrant back out of the surface discontinuity, producing an indication which fluoresces under ultraviolet lighting.

19

The 1989 Sioux City Crash

On July 19, 1989, United Airlines Flight 232 crashed while attempting to land at Sioux City airport. Of the 296 persons on board, 111 died and 185 survived, some literally walking out of the shattered aircraft into Iowa cornfields. Miraculously, thirteen passengers received no injuries at all.

Investigation of the crash revealed that the tail engine of the DC-10 had exploded in flight and that shrapnel from the engine had severed all hydraulic lines in the tail, allowing the fluid to leak out. As a result, all hydraulically powered controls—which includes the flying surfaces on the wings and tail—no longer functioned. The only way the flight crew could control the aircraft was by increasing or reducing power to the two wing engines. This mode of control causes the aircraft to rise and fall like a boat encountering large waves. It is something of a miracle that pilot Alfred Haynes and trainer pilot Dennis Fitch brought the plane as close to a landing as they did. As a McDonnell Douglas test pilot pointed out, flying the plane without essential controls "is not a trainable maneuver." No pilots were able to land a similarly disabled aircraft in simulator tests.[1]

The Sioux City crash raises a number of new questions about the DC-10. First, did McDonnell Douglas adequately design the control system in the tail of the aircraft, given the dangers of a tail engine design? Second, are the policies and institutional arrangements for protecting against this kind of catastrophic engine failure as good as they should be? Here we encounter again the familiar tradeoff between safety and economy which exhibits some unusual features because of the institutional arrangements of air safety regulation. The answers to these questions depend upon seeing how various technical features of aircraft design enter into them.

HYDRAULIC SYSTEM DESIGN

The DC-10's three separate hydraulic systems are carried in one-half inch stainless steel tubing at a pressure of 3000 pounds per square inch. Those lines converge and run close together in the tail of the DC-10 as they connect with the horizontal stabilizers and rudder. Despite the vulnerability of hydraulic lines in this arrangement, McDonnell Douglas did not include any type of shut-off valve or additional shielding in the design of the hydraulic system. In contrast, Lockheed's L-1011, which is the only other jumbo jet to have a driving engine[2] in the tail, has a shut-off valve to protect against loss of hydraulic fluid in the event of explosive failure of the tail engine.[3] In addition, both the L-1011 and the Boeing 747 have a fourth hydraulic system which provides extra redundancy.

The value of this additional hydraulic system was dramatically illustrated on September 22, 1981. The tail engine of an Eastern Airlines L-1011 completely disintegrated as the plane reached 10,000 feet after taking off from Newark, New Jersey. There was extensive damage, and three of its four hydraulic systems were lost. The fourth system was damaged but was able to function, allowing the pilot to land safely.[4]

Other tail engine jets, such as the familiar Boeing 727 with its three tail-mounted engines, have mechanical back-up systems which use wires and pulleys. The flying surfaces on the jumbo jets are very large and require so much force to move them that mechanical operation, even as a back-up system, is not practical.

Although the Boeing 747 does not have a driving engine in the tail, a faulty repair to its aft pressure bulkhead in 1985 resulted in explosive decompression which damaged the tail and caused loss of all hydraulic controls. The Japan Air Lines flight was not as lucky as

Sioux City; the plane flew erratically for a short time and finally crashed, killing 531. The FAA required Boeing to install check valves after that crash.

UNCONTAINED ENGINE FAILURE

The cause of the tail engine's disintegration in the Sioux City crash was a flaw in the stage-one fan disk, which is attached to the engine shaft and holds the turbine blades. The flaw was a metallurgical anomaly called "hard alpha" which allowed a crack to begin which eventually grew large enough to cause disintegration of the disk. The anomaly was in a part of the disk where it is difficult to detect cracks.[5]

The fan disk is three feet in diameter and weighs about 370 pounds, so when it cracked and disintegrated the DC-10 experienced "uncontained engine failure," in which parts of the engine escape from its protective cowling (see fig. 19.1). The National Transportation Safety Board (NTSB) "recognizes that current certification rules view engine failures that result in the liberation of high-energy rotating parts is an intolerable event for which total protection cannot be practically provided."[6] Shielding capable of retaining a 370-pound disk which disintegrates while rotating at high speed would have to be extremely heavy and hence impractical. This means that the large fan jets that power the new aircraft pose a serious safety threat to the integrity of the aircraft.

Uncontained engine failures are not unusual. At least sixty-seven have occurred on commercial aircraft since 1983. In a typical accident, blades from the turbine break off the fan disk and are hurled out the front of the engine or through its protective cowling. This poses obvious dangers to hydraulic lines and other structural components in the vicinity.

During a training flight in May, 1972 the tail engine of a Continental DC-10 broke apart. Most of the debris flew out the rear of the engine and caused little damage. An uncontained wing engine failure on a 1977 UTA flight resulted in an uncommanded slat retraction when hydraulic and control cables were damaged. Like the Chicago crash, this incident occurred just after takeoff, but the plane was able to land safety. A similar incident took place when a wing engine on an Air Florida DC-10 disintegrated while the plane was accelerating to take-off speed. The pilot was able to safely abort the takeoff, but examination of the wing revealed damage to control sys-

FIG. 19.1

tems and an uncommanded slat retraction. Had this flight been further into its takeoff roll, it could easily have been a replay of the Chicago crash of 1979.

A bizarre example of the dangers of blade loss occurred in November, 1973, when a DC-10 wing engine disintegrated and shattered a window in the passenger cabin. The passenger sitting by the window was sucked out of the aircraft near Albuquerque and was never found. It was the first DC-10 fatality.

The engine failures that caused the Sioux City crash and the 1981 Newark incident were not simply the loss of turbine blades. In both cases the large disk that holds the blades also came apart, releasing much larger and more destructive pieces of metal. Although accidents of this destructiveness are relatively infrequent, they highlight the danger of locating a driving engine in the tail of the aircraft.

THE INHERENT DANGER OF TAIL ENGINES

Tail engines are inherently more dangerous because of the proximity of hydraulic lines and control cables to the engine. All control systems converge in the tail and are more vulnerable to multiple damage. In addition, the tail is critical to aircraft control. While it is possible to fly a DC-10 without rudder controls, the position of the horizontal stabilizer is critical. Jumbo jets are inherently nose-heavy, so that the horizontal stabilizer is needed to keep the nose of the aircraft up. Loss of rear elevator control is very serious and can easily result in loss of the aircraft.

From an ethical standpoint, greater danger entails a greater obligation to take precautions to prevent harm.[7] Lockheed's L-1011, the only other jumbo jet with a rear-driving engine, installed both an additional hydraulic system and a check valve. In contrast, McDonnell Douglas used only three hydraulic systems and has installed check valves in both the wing and tail only after being directed to do so by airworthiness directives issued by the FAA. Both directives were the result of serious accidents, Chicago (1979) and Sioux City (1989).

In addition, McDonnell Douglas was the only manufacturer of jumbo jets that did not have any form of positive mechanical locking device for its slats. Such a locking device probably would have prevented the Chicago crash. As with the hydraulic check valves, only FAA action (and that only after a fatal crash) prompted McDonnell Douglas to install locking devices on the slats.

TESTING FOR FAULTY PARTS

There are two concepts in aircraft design that are used to insure its safety. *Fail-safe* means that the failure of a component will not cause fatal damage to critical parts of the aircraft. The redundancy of hydraulic systems is an example of this concept. If one system fails the aircraft can be flown with the others. Similarly, critical controls, such as the horizontal stabilizers, are divided into different segments so that failure of one hydraulic system means that only a portion of the flying surface is lost. If the entire horizontal stabilizer was powered by one hydraulic system, there would be no redundancy to allow that system to fail safely.

Items such as fan disks, whose failure is very dangerous, are periodically replaced to prevent failure. This approach emphasizes the *safe life* of the component rather than protection against its failure mode. Similarly, engines are serviced and critical parts replaced on a safe-life basis rather than waiting until failures occur. These concepts are not mutually exclusive. Even though engine parts are replaced on a safe-life basis, their failure should not jeopardize the safety of the aircraft. Redundancy built into control systems, for example, is supposed to take care of such situations.

Some components, such as the wings, cannot fail safely, nor can they simply be replaced, like fan assembly disks. As a result, these items are designed to carry much more than the maximum feasible load. In the Boeing manufacturing plant there is a photograph of a wing being tested. In the test the wing is bent far upwards above the fuselage. Like bridges, the wings are designed to withstand loads much greater than anything anticipated in actual service.

METAL FATIGUE AND SKIN FAILURE

The skin of an aircraft is a structural component. In a large building, such as a skyscraper, the skin (the façade) is simply hung on a frame which carries the load (weight) of the building. But the skin of a modern jet aircraft is a load-bearing structure: there is no frame underneath to do that work. Consequently, any rupture of the skin would be very serious, particularly since this could cause explosive decompression of the interior of the aircraft.

The skin of a jet aircraft is also subject to a special kind of stress when the plane is pressurized. In order for passengers to breathe comfortably in high altitudes, the air in the cabin is pressurized. As the plane increases its altitude, the outside air pressure falls and cabin pressure must be increased. This results in a load of five to ten pounds per square inch on the skin of the aircraft. In effect, the airplane is blown up like a balloon when it is pressurized and then deflated when it reaches a lower altitude. This process of pressurization and depressurization is called a *cycle*. Every cycle causes the skin to flex, and like repeated bending of a paper clip, this results in metal fatigue and the potential for failure. The most dramatic example of this process occurred in the 1950s, when British de Havilland Comet jets began to break apart in flight not long after they were put into service. Investigation revealed that flexing of the wings had caused metal fatigue where they were joined to the fuselage. The fatigue was so great that, like the paper clip, they simply snapped off.[8]

A more recent example of metal fatigue failure took place in 1988 when an Aloha Airlines Boeing 737 suddenly lost eighteen feet of the top portion of the aircraft behind the cockpit. Miraculously, the plane did not break apart and the pilot was able to land safely, although one flight attendant was swept to her death and many passengers were injured. Incidents such as this have generated a debate about the safety of aging aircraft. As aircraft experience more and more cycles of pressurization, fatigue cracks develop in the skin of the aircraft. The older the aircraft, the bigger and more numerous the cracks. Should there be a "safe-life" limit on the aircraft itself? Or can a program of inspection assure safety of older aircraft? Operators clearly would prefer to inspect and keep functioning aircraft in service. And any forced sale of aging aircraft, most likely to third-world countries, raises difficult ethical issues. Should we allow the sale of aircraft deemed no longer safe to other countries where it will be used to carry passengers?

INSPECTION AND TESTING FOR CRACKS

Metal fatigue in aging aircraft and the failure of the fan assembly disk in the Sioux City crash are related problems. Both involve metal failures that can be very serious for the safety of the aircraft. And both are prevented by a program of inspection using a variety of tests to detect flaws or cracks before they become dangerous.

United Airlines was using a fluorescent dye test for its periodic inspection of the fan assembly disk. In this procedure a special dye is applied to the part and it is inspected under fluorescent light. Tiny cracks or other flaws will retain some of the fluid and will stand out. United Airlines FAA-approved maintenance procedures require that whenever the engine is dissembled this test is to be applied. The engine that exploded was tested in this manner 760 cycles before the accident. The NTSB concluded that there is "evidence suggesting that, at the time of this fluorescent penetrant inspection, the crack was of a detectable size. United Airlines believes that stresses associated with the shot peened surface of the disk may have kept the crack closed, making it undetectable, or caused the surface length to be very small."[9] This was the only test required by the FAA prior to the Sioux City crash. The more sensitive ultrasound test, in which the disk is immersed in water and subjected to sonic waves, was neither required nor used. It is generally agreed that this test would have detected the crack which had been propagated from a flaw in the casting of the disk.

Inspection is an important feature of aircraft safety. A substantial portion of NTSB recommendations and FAA airworthiness directives concern inspection of critical parts for wear or signs of premature failure. Sometimes the inspection is only visual, but it may also require the use of crack detection tests. Some relatively simple tests involve the application of a liquid dye like the fluorescent dye test. More elaborate tests include ultrasound or x-ray analysis. These are nondestructive tests, although some require disassembly of the part to be tested. Eddy current testing uses an electrical current to reveal cracks, but it may require the removal of rivets to check for skin cracks.

SAFETY AND ECONOMY

Tests and inspections involve a tradeoff between economy and safety. Even if a component is easily visible, inspections require time and effort. If, in addition, the inspection requires disassembly and/or destruction and replacement of rivets, the cost to the airline goes up rapidly. The familiar conflict between the FAA's obligation to insure safety and to promote the aviation industry is part of the Sioux City crash as well.

The FAA is reluctant to require design changes or to order replacement of expensive parts unless there is substantial evidence that it is necessary. Critics charge that this philosophy is one that reflects the interests of industry: it is cheaper to inspect than to replace or redesign. In addition, by placing the burden of proof on the claim that a design is not adequately safe, the FAA has decided to err on the side of economics rather than safety. The result is that often the substantial evidence that a problem needs more than inspection turns out to be a fatal crash.

SAFETY AND THE DC-10

All three crashes—Paris (1974), Chicago (1979), and Sioux City (1989)—were caused by failures that should not have disabled the aircraft. Loss of a cargo door, an engine, or an engine disintegration should not render the DC-10 unflyable. It is only because the inadequately protected control system of the DC-10 failed in these circumstances that 730 people died.

The designers of the DC-10 did not successfully make the jump to the jumbo jet with respect to protecting its control system from collapsing under ordinary failures. Douglas had never built a jumbo jet or an aircraft with a driving engine positioned in the tail. Both of these new tasks were undertaken in a competitive race with Lockheed to produce the second jumbo jet for the aviation market, knowing there was not room for three at that time. It is hard to escape the conclusion that Douglas chose to build an airplane with a relatively thin depth of safety defense in order to save time and keep costs down. This does not mean that Douglas designers necessarily thought they were building an inadequately safe airplane (although some certainly did, as the Applegate memo reveals). The safety of the airplane is primarily based on the depth of its defenses against serious accidents, and the DC-10's was simply not deep enough. Douglas thought they could get by with fewer protections than were installed on the Lockheed L-1011 and the Boeing 747, but events have clearly shown that they were wrong.

NOTES

1. *The Washington Post*, November 1, 1989, p. A3.

2. Boeing has an auxiliary power unit (APU) in the tail which is used to provide electrical power, but it is small and presents little hazard to hydraulic lines.

3. *The Wall Street Journal*, July 24, 1989, pp. 1, A4.

4. NTSB Safety Recommendation A-89-95 through -97, August 17, 1989, p. 3.

5. National Transportation Safety Board Safety Recommendation A-90-88 through -91, (June 10, 1990), pp. 1-2.

6. NTSB Safety Recommendation A-89-95 through -97, August 17, 1989, p. 3.

7. For a discussion of this point in connection with the DC-10, see Martin Curd and Larry May, *Professional Responsibility for Harmful Actions* (Dubuque, IA: Kendall-Hunt Publishing Co., 1984). This idea also appears in law. In *U.S. vs. Carroll Towing Co.* (159 F 2nd 169 (1947)) a barge operator was held responsible for damages because the procedures to keep the barges from breaking loose were not consonant with the severe damage that could occur.

8. See Stanley Stewart, *Great Air Disasters* (London: Arrow Books, 1988), pp. 54-90.

9. Op. Cit., p. 3.

THE AVIATION SAFETY SYSTEM

20

Statement of Ralph Nader*

STATEMENT RALPH NADER
Subcommittee on Government Activities and Transportation
of the
House Committee on Government Operations
Washington, D.C.
June 18, 1979

Mr. Chairman, Members of the Subcommittee: thank you for your invitation to testify this afternoon.

Ever since American Airlines Flight 191 crashed on takeoff at O'Hare International Airport on May 25, there has been growing public concern about how well the Federal Aviation Administration is protecting the air-traveling public. When a disaster of the magni-

---

* Reprinted, with editorial changes, from *FAA Certification Process*, Hearings before the Subcommittee of the Committee on Government Operations, House of Representatives, Ninety-Sixth Congress, June 11, 18; October 9 and 10, 1979.

tude of the O'Hare crash takes place—especially when it happens only eight months after the midair collision over San Diego—the public has a right to be outraged and to demand that the federal government take swift and effective action to improve air safety.

In the wake of this tragedy, the key question is, "Who is protecting the traveling public?" Ordinarily, one might assume that the FAA is doing the job, since that is the mission assigned to it by statute. But as these hearings unfold, we see an all-too-familiar pattern emerging, namely, that the FAA is part of the problem, not the answer. It has been an agency heavily indentured to the airlines and the manufacturers, an agency which is chronically unable to anticipate safety problems, unwilling to acknowledge that hazards exist, and incapable of swift and effective action to remedy dangerous conditions.

FAA administrators come and FAA administrators go, but the basic problems persist. That is why the public must look to Congress for solutions. It is Congress which established the FAA, Congress which defined its mission, and Congress which must act since the FAA has consistently failed to insure the higher level of safety which the public demands and expects.

It is this issue which I will address today. The FAA's actions—and omissions—in connection with the Chicago crash are indicative of problems which have long been apparent, and which have been examined by this subcommittee in a number of oversight hearings in recent years. I hope that you will use that record and the record from these hearings as the basis for comprehensive reforms in the way the FAA does business.

The FAA's response to the DC-10 crash at O'Hare is symptomatic of its reluctance to acknowledge the existence of safety problems which might be expensive to the industry or which call for strong and urgent action. Several hours after the May 25th crash, I called upon FAA administrator Langhorne Bond to ground all DC-10s in U.S. fleets until they were inspected and certified as safe. The nature of the O'Hare accident indicated that there was some serious design or structural problem with the DC-10. After all, engines are not supposed to fall off of airplanes, but even if they do, catastrophic loss of control should not result. I was particularly concerned because of prior DC-10 accidents where safety deficiencies could be traced back to the certification process. Immediate precautionary action was necessary.

Mr. Bond felt no need for urgent action, however, and the next morning, May 26, he rejected the possible grounding of DC-10s as "premature."

On Sunday, May 27, I wrote to Secretary of Transportation Brock Adams, urging him to overrule Mr. Bond. That letter described previous DC-10 accidents which are well known to this subcommittee and to the public. The most devastating example, of course, is the Turkish Airlines DC-10 crash near Paris on March 3, 1974. Shortly after takeoff, a cargo door blew off and the explosive decompression resulted in a crash, killing 346 people. Between 1971 and 1974, the General Electric CF-6 engine which powers the DC-10 failed dozens of times because it was unable to stand up under icing conditions. On March 1, 1978, two people died and thirty-one were seriously injured after tires blew out as a Continental DC-10 attempted to take off from Los Angeles International Airport. The NTSB report noted that tire strength standards were last set in 1962 and did not take into account actual operating characteristics of wide-bodies such as the DC-10.

Did the FAA take strong and decisive action? Hardly. On May 28, the discovery of a thrust link bolt prompted the FAA to issue an airworthiness directive which grounded the DC-10s temporarily until the bolts and pylon aft bulkhead could be visually inspected.

The inadequacy of the limited inspection was soon demonstrated.

- The first round of inspections on May 28 turned up what Mr. Bond called "grave and potentially dangerous deficiencies in many of the pylon mountings." He issued a slightly broader directive on May 29. Even after issuing this second directive, Mr. Bond refused to order a more comprehensive inspection because no one could tell him what new defects a thorough inspection might uncover. This is similar to a doctor being told that a patient suffers from heart and liver trouble but refusing to conduct a thorough physical because no one told him anything else was wrong.
- As reports of cracks in the pylon area continued to pour in over the next few days, National Transportation Safety Board officials urged immediate action. NTSB member Philip A. Hogue publicly suggested that the FAA orders did not go far enough, and the Safety Board on June 4 recommended immediate inspection of all DC-10s whose engine pylon assembly had been removed and reinstalled. The NTSB also suggested that maintenance procedures which were at odds with the McDonnell Douglas recommended procedures might have played a role in causing the cracked pylon assemblies.

The FAA on June 4 issued its third grounding order, this one for planes whose pylons had been previously removed and reinstalled. Inspection of two American DC-10s which had previously passed inspection led to the discovery of new cracks on June 5, and this precipitated Mr. Bond's suspension of the DC-10's type certificate on June 6. This was twelve days after the O'Hare crash.

It is appalling that it took twelve days after the O'Hare crash for the FAA to realize that the DC-10, in its words, "may not be of proper design, material, specification, construction, and performance for safe operations, or meet the minimum standards, rules or regulations prescribed by the Administrator." When an engine falls off and catastrophe results, isn't defective design the most logical hypothesis, particularly in light of the DC-10's prior record? Yet the traveling public was exposed to unnecessary risks for twelve days due to the FAA's timid, stop-and-start approach to grounding the planes. The FAA's failure to take strong steps immediately after the crash did nothing to reassure the public that the agency put safety first. Its behavior will make it harder for the public to be confident that the DC-10 is safe once the planes are flying again.

FAILURE TO SET ADEQUATE SAFETY STANDARDS

The reluctance to take decisive steps to promote aviation safety in the wake of the nation's worst air disaster is typical of many other substantive failures the FAA has exhibited. I have already mentioned the deficiencies which took place during the FAA's certification of the DC-10. Let me cite some other areas where FAA obstinacy and delay have needlessly endangered the traveling public, or where action was taken only after a fatal crash.

- It was not until last year's San Diego midair collision that the FAA proposed a comprehensive review and upgrading of the nation's air traffic control system.
- It was not until two tires blew out on the Continental DC-10 last year that the FAA proposed upgrading tire standards for the first time since 1962.
- It was not until a TWA jet flew into a mountain near Dulles Airport in December 1974 that the FAA, under intense congressional pressure, required airlines to install Ground Proximity Warning Systems (GPWS) to let pilots know when they are sinking too fast or approaching mountainous terrain.

- One subject on which the subcommittee has conducted several hearings is the FAA's failure to develop a workable Collision Avoidance System. An October 1974 GAO report noted that the FAA had been studying the midair collision problem for almost two decades; even after the disaster at San Diego last September, the FAA is not planning implementation of a national standard until the mid-1980s, and the Air Line Pilots Association has seriously questioned whether the system the FAA favors will be the most effective.
- Upgraded safety standards for commuter airlines is another area where it took years for the FAA to act. In 1972, the NTSB called for a complete and expedited revision of Part 135 of the Federal Aviation Regulations, which governs the operations of some 200 commuter airlines in this country. FAA rule-making dragged along with no signs of progress until 1976, when the Aviation Consumer Action Project published a report which concluded that (1) commuters as a whole had an accident record three times worse than trunklines, and (2) the higher rate is attributable to differences between safety regulations governing the two types of carriers. This report spurred the FAA to step up its work, but it was not until October 1978 that final rules were issued, and these new regulations still allow discrepancies between standards governing air carriers and those applicable to commuter carriers.
- The problem of survivability in aircraft crashes provides another enduring example of FAA "standards-avoidance behavior" in the face of a serious danger. In many take-off and landing accidents, the crash impact forces are low and passengers often can survive the crash itself. People are often seriously injured, however, when their seats collapse or break loose, hurling them about the cabin and injuring passengers or blocking emergency evacuation exists. Comprehensive seat strength standards were set in 1952, and except for one minor revision in 1957, they have remained unaltered for the past twenty-seven years. The crash forces which an automobile seat must withstand are considerably higher. Incredible as it may sound, aircraft which are about to be certified for use in the 1980s and 1990s will only have to meet the seat strength standard devised for aircraft of the early 1950s. In October 1977, Aviation Consumer Action Project, the Association of Flight Attendants and the Flight and Engineers International Association petitioned the FAA to upgrade these rules and to require dynamic testing of passenger and crew seats. After twenty months, they have not received a formal response.

Burning cabin interior materials which emit smoke and toxic gases have led to dozens of passenger deaths in supposedly "survivable" crashes, but the problem has received low-priority treatment for over a decade. The FAA has had flammability standards for 30 years, but in many crashes, there are more hazards than just those resulting from fire and high levels of heat.

In many survivable accidents, passengers die not from the crash impact or from fire but from the inhalation of smoke and carbon monoxide or hydrogen cyanide which is created during combustion of draperies, upholstery, seat cushions, ceiling panels, and other cabin interior materials. The FAA began research on this problem in 1966 and put out a rule-making notice on smoke emissions in 1969 and 1974, and a notice on toxic gas emissions in 1975. However, the FAA withdrew these notices in August 1978 and assigned the job of finding a solution to the cabin materials problem and to the related fuel fire problem to a special advisory committee. Administrator Bond has now told Congress that he is committed to swift action on the cabin materials problem and has charged an industry-dominated advisory committee with making recommendations by October 1 as to what short-term steps can be taken. However, it took thirteen years to get from preliminary research to formation of an advisory committee, and a healthy dose of skepticism is necessary.

These are only some of the areas where the FAA has been derelict in setting minimum safety standards, and I understand that the General Accounting Office is conducting a thorough examination of how the agency has perceived and responded to twenty-six specific aviation hazards. Its findings should be particularly helpful to this subcommittee.

FAILURES IN THE INSPECTION PROCESS

Thus far, I have concentrated on FAA deficiencies in setting safety standards. There is another important area where the FAA has abdicated its responsibility to protect the traveling public, namely its failure to monitor airline safety practices in a thorough, persistent fashion.

In September 1977, this subcommittee held hearings on "Airline Deregulation and Aviation Safety" which revealed just how heavily the FAA's "inspection" process relies on industry self-monitoring. Testimony from pilots and airlines indicated that the FAA's oversight of maintenance procedures is carried out primarily by monitoring maintenance logs. As one Western Airlines representative testified, "The FAA's participation occurs, in the main, only after a difficulty has occurred, in an effort to determine whether we had in fact operated in accordance with safety standards we had set."

The whole inspection process has traditionally been shrouded in secrecy. From 1966 to 1977, a center piece of the FAA's surveillance program was the "Systemworthiness Analysis Program," known in the industry as "SWAP." SWAP is essentially a system of periodic audits of the airlines' maintenance and operational procedures, and it was performed by teams of FAA inspectors and auditors on a twelve to eighteen month cycle. The inspectors' initial findings and recommendations are disclosed to airline management, and a final SWAP report is thereafter prepared. These reports contain the ultimate findings and recommendations of the SWAP teams. Although, SWAP reports are a critical gauge of both airline operations and FAA oversight functions, they have never been disclosed by the agency, despite a nine-year effort by public interest groups.

In the summer 1970, Reuben Robertson and Jerome Simandle of the Center for Study of Responsive Law demanded that SWAP reports be released under the Freedom of Information Act. The FAA refused at the request of the Air Transport Association. FAA argued that release would have a "chilling effect" on the airlines' compliance with FAA inspection procedures and the airlines' willingness to be forthcoming with SWAP inspectors. The FAA administrator blocked disclosure under section 1104 of the Federal Aviation Act (49 U.S.C. § 1504) which allows the withholding of information whose disclosure is not, in the administrator's judgment, required "in the interest of the public."

We disagreed with the FAA's interpretation, and took them to court in 1971. The public has a right to know the adequacy of FAA surveillance of airlines' maintenance and operations practices. Congress gave the FAA broad enforcement powers, and the FAA does not have to rely on industry good will to fulfill its statutory mandate to protect the public.

The District Court and the Court of Appeals ordered the release of the SWAP reports, but the Supreme Court reversed in *Administrator, FAA v. Robertson*, 422 U.S. 255 (1972). In an opinion by Chief Justice Burger, the Court held that the Freedom of Information Act did not override section 1104 and that the administrator had the right to decide what information is to be suppressed "in the public interest."

However, Congress responded in 1976 during its consideration of the government in the Sunshine Act, when it included a provision to amend the Freedom of Information Act and specifically overrule the *Robertson* decision.

Because Congress had made it clear that the FAA had no basis for refusing to release SWAP reports, the Aviation Consumer Action Project and the Aircraft Mechanics Fraternal Association requested access to SWAP reports in September 1976. The FAA again refused, and ACAP and AMFA were thus forced to file suit in March of 1977. Almost nine months later, the FAA decided that nothing in the Freedom of Information Act prevented release of SWAP reports.

At this point, however, two airline trade associations—the Air Transport Association and the National Air Carrier Association— intervened and tried to permanently enjoin release of the SWAP reports, claiming everything from the law of defamation to five Freedom of Information Act exemptions previously rejected by the FAA.

The District Court rejected the majority of the airline claims, and last November, we sought to obtain a final ruling requiring disclosure of SWAP reports. In order to expedite release, we agreed to leave for another day questions concerning materials claimed to be exempt. Last week, in light of the DC-10 crash and the significance of the findings of the SWAP teams to the adequacy of the FAA inspection program, we filed papers asking the U.S. District Court to expedite a final ruling in our case.

What is deplorable about the whole SWAP litigation is that the FAA and the airlines between them have managed for nine years to prevent the public from learning just how well or how poorly the FAA monitors airline maintenance and operations.

Even more disturbing is the FAA's response to the 1976 law which was intended, in effect, to open SWAP reports to public scrutiny. Shortly after ACAP and AMFA filed suit in 1977, the FAA revised the SWAP program so that inspections would no longer be conducted every twelve to eighteen months. Instead, they would be performed on an "as needed" basis, and the "need" for an inspection would be determined by FAA tracking of the daily Mechanical Reports and Service Difficulty Reports. If they spot a trend, the regional office or district office might dispatch a team to inspect a particular airline for that particular problem, but as one FAA employee conceded to us, it was possible under this new system for an airline to avoid a thorough inspection for two years.

How can Congress condone an agency which refuses to honor the Freedom of Information Act and the Sunshine Act, and worse, which refuses to exercise its enforcement powers for fear of offending its friends in the airline industry? Congress gave the FAA broad enforcement powers and it meant the FAA to exercise those powers. Nowhere in the Federal Aviation Act did Congress condition the FAA's use of its authority on airlines' approval.

A NTSB safety recommendation issued on May 9, 1979, provides a rare glimpse of how FAA enforcement procedures work in practice. The case involved a commuter carrier, Antilles Air Boats, one of whose flights crashed in the Virgin Islands on September 2, 1978, killing the pilot and three of the ten passengers on board. The Safety Board investigation revealed poor operational and maintenance practices, falsified logbooks, and management practices which often condoned or encouraged violation of federal regulations. Even worse, the company was a repeat offender—FAA enforcement letters written in 1977 and 1978 noted deficiencies which had been exposed in a 1975 SWAP inspection.

The Safety Board noted that two inspectors were assigned to the Flight Standards District Office in San Juan, where Antilles was based, and that these inspectors were conscientious, but that their suggestions were not acted upon by higher levels in the FAA. As the Safety Board found, "The results of inspections, the numbers of enforcement actions, and the accident/incident record should have demanded immediate corrective action by the FAA. Instead, the Safety Board discovered that compromises of violation action were so common that the enforcement program was rendered ineffective." The NTSB has been urging more and better surveillance over smaller carriers such as Antilles since 1972, but thus far the advice has fallen on closed ears.

SUGGESTIONS FOR REFORM

This is a thumbnail sketch of some of the FAA's grossly inadequate responses to its duty to protect the traveling public. Many of these tragedies and problems are familiar to the subcommittee, and in fact, much of the record on some of these problems was developed in your earlier hearings. But the issues are worth considering again, because Congress must move beyond the hearings stage to the action stage. Having built the record which you have, the burden is on this subcommittee not just to oversee, but to overhaul this sluggish dinosaur of an agency. Let me suggest some areas where improvement can be made:

1. Make the FAA first and foremost a safety agency. The Federal
   Aviation Act now gives the agency a schizophrenic mission. On the
   one hand, it is supposed to insure the "highest possible degree of
   safety" (section 601, 49 U.S.C. § 1421). On the other hand, it is
   supposed to "encourage and foster the development of civil aeronautics" (section 305, 49 U.S.C. § 1346). Even if Mr. Bond sees no
   conflict between these two goals, the FAA has in the past weighed
   human safety against corporate profits, and come down in favor of
   the latter. Witness the infamous 1972 "gentleman's agreement"
   where Douglas agreed to fix the DC-10 cargo door if the FAA
   agreed not to put out a public airworthiness directive that might
   scare off potential DC-10 buyers. Or consider the years of foot-
   dragging before a TWA jet crashed into a mountain near Dulles
   over four years ago. A Ground Proximity Warning System could
   have helped to prevent that accident, but the FAA did not order
   installation because it might cost the airlines some money.
       When you consider that more than seventy-five years have
   passed since the Wright brothers' first flight, it is hard to see why
   giant airline and manufacturing companies need this type of governmental "promotion." They are perfectly capable of promoting
   civil aeronautics on their own.
2. Require the FAA to put major emphasis on research and development activities, on issuing timely safety standards, and on constantly reviewing those standards to make sure they are up to date.
   We are now twenty years into the jet age, ten years into the jumbo-
   jet era and on the threshold of certifying a new generation of aircraft—such as the Boeing 757 and 767—which will be flying past
   the year 2000. Yet the FAA is holding manufacturers to 1952 seat
   strength standards, 1962 tire standards and no standards for emissions of smoke and toxic gases by cabin interior materials. When
   you consider the ongoing changes and development in aviation technology, the FAA cannot afford to lag behind, and it should be setting
   standards which keep up with the times. Congress should directly
   set for the FAA deadlines for assuring these standards, as was done
   for auto fuel tank standards. A generation of chronic delay via the
   agency's legendary apathy and connivance is enough.
3. Set target dates for responding to rulemaking petitions and for
   concluding rule-making proceedings. The U.S. Court of Appears
   has observed that

   > nine years should be enough time for any federal agency to
   > decide almost any issue. There comes a point when relegating

issues to proceedings that go on without conclusion in any kind of reasonable time frame is tantamount to refusing to address the issues at all—and the result is a denial of justice. [*Nader v. FCC*, 520 F.2d 182, 206 (D.C. Cir. 1975)]

Unfortunately, the FAA has shown that it can easily take longer than nine years to deal with critical safety needs and still fall far short of taking effective action. That is why it is essential to establish deadlines and goals. The Civil Aeronautics Board has a policy of responding to rule-making petitions within 120 days after answers have been filed, but the FAA has rejected even this modest proposal. As a result, the public's petitions sit on the FAA's docket for years before the agency deigns to act. Moreover, even when they initiate a rule-making proceeding, there is no guarantee that it will be rapidly concluded. That is why it is essential for there to be deadlines for the completion of each phase of a proceeding, with public explanations and sanctions if target dates cannot be met.

4. Upgrade FAA practices for regulating airline maintenance procedures, such as increasing the number of field staff to perform this oversight. As a September 1978 article in *Air Transport World* points out, "whatever can be accomplished by excellent design, poor maintenance can undo." In December of 1969, Reuben Robertson of the Center for Study of Responsive Law prepared a comprehensive report detailing the inadequacies of FAA practices in this area. It is shocking to realize a decade later that the conclusions of that report are still timely.

5. Require FAA personnel to conduct periodic inspections, which are unannounced and thorough. The inspection team's findings and recommendations should be publicly available, and the law should provide adequate penalties for violations of any statutes or regulations by industry or agency personnel. Congress should also consider improving diligence and candor by inspectors by establishing incentives for deficiencies uncovered. The history of government inspectors, as in the meat and poultry area, reveals pressure by industry and the inspectors' supervisors to look the other way or be transferred. There needs to be a counteracting measure.

6. Establish personal accountability procedures which can be invoked by aggrieved parties with respect to action by FAA officials. Excessive security of tenure leads to excessive irresponsibility. FAA officials must have something to lose if they misbehave and the law must provide that possibility.

7. Establish a program which encourages and facilitates citizen and consumer group participation in the formulation of safety standards, and include in that program provisions for compensating the participation costs of such parties.
8. Reform the aircraft certification process and end the practice of delegating responsibility for checking out new aircraft to company employees.
9. Consider splitting the FAA's regulatory functions from its operational functions. At present, the FAA has enormous responsibility not only for the development and enforcement of safety standards, but also for the daily operation of the air traffic control system and the funding of airport construction and navigational facilities and equipment from the Airport Trust Fund. It may be that the two functions could be better performed by separate agencies.

These suggestions are offered to spur the Congress to take positive steps which will lead to permanent improvements in the way the federal government regulates the safety of air travel. The time for comprehensive reform measures was years ago.

Thank you.

21

Aviation Safety: Management Improvement Needed in FAA's Airworthiness Directive Program*

This report responds to an October 11,1988, request that we explore, through several assignments, the Federal Aviation Administration's (FAA) ability to provide meaningful oversight of aviation safety. Pursuant to your request we previously reported on FAA's removal procedures for pilot examiners,[1] the agency's inspection management system,[2] and changes under consideration regarding aviation medical standards.[3] As agreed with both subcommittee staffs, this report focuses on FAA's oversight of the airworthiness directive (AD) program because ADs are key elements of FAA's safety responsibility. ADs are rules that FAA issues requiring airlines to correct conditions in their aircraft, such as cracking and corrosion, that can

---

* Reprinted with editorial changes from the United States General Accounting Office Report: *Aviation Safety: Management Improvement Needed in FAA's Airworthiness Directive Program*, 1990.

jeopardize safety. Because of the AD program's critical nature, we agreed to examine whether (1) FAA's oversight is adequate to determine airline compliance with AD requirements and (2) safety-related information maintained by FAA could be used to improve the program's effectiveness. Our review focused on FAA's oversight of the AD process, not on the merits or validity of individual ADs.

RESULTS IN BRIEF

FAA's oversight of the AD program is inadequate to determine whether the airlines are complying with ADs. The National Transportation Safety Board's (NTSB) investigation of the widely publicized April 28,1988, Aloha accident, in which the aircraft lost the upper portion of its cabin while in flight, and other FAA special inspections have documented several important examples of AD noncompliance. FAA does not know the extent of AD compliance throughout the airline industry because FAA inspectors do not, and are not required to, verify AD compliance during each inspection. In addition, when AD compliance is checked, current guidance requires inspectors to report only noncompliance. As a result, FAA does not have information on the extent to which airlines are in compliance with ADs. FAA needs to measure the extent of noncompliance and decide what actions it should take to ensure AD compliance throughout the airline industry.

FAA inspection guidelines are very broad, in that they allow inspectors to decide whether to check for AD compliance, which ADs to include in their check, and which aircraft to inspect. FAA inspectors do not routinely check for AD compliance during airline inspections. For example, during our review we accompanied FAA inspectors on ten routine surveillance inspections and noted that AD compliance was checked in only two instances because the inspector decided to emphasize non-AD-related items.

In addition to obtaining more complete and comprehensive information from its safety inspectors, FAA should better use its existing safety data bases. Use of such data could help focus airline inspections on problem areas, thereby improving the effectiveness and efficiency of FAA's limited inspector work force.

BACKGROUND

The Federal Aviation Act of 1958, as amended (P.L. 85-726), established the safe travel of U.S. air passengers—about 492 million in 1988—as a joint responsibility of FAA and the airlines. FAA promotes aviation safety by issuing regulations that stipulate certain requirements that airlines must meet to operate commercial aircraft. FAA's aviation inspectors then monitor to ensure that airlines comply with these safety requirements. When airline accidents do occur, mechanical failures—such as in landing gear and aircraft structural components—are factors in forty percent of the accidents. FAA's AD program addresses such unsafe conditions and is a key element of FAA's safety responsibility.

Although FAA certifies new aircraft models as safe before they are used in commercial service, it also issues ADs to address unsafe mechanical conditions that surface after the aircraft has been in use. ADs are requirements that FAA issues for the airlines to identify and correct unsafe aircraft conditions that have occurred, or are likely to occur, in other aircraft of the same design. ADs also prescribe corrective actions that airlines must take to correct identified problems in their aircraft. For example, a 1988 AD identified several incidents involving Boeing 737s in which an engine bolt failed. In one case, the engine separated from the aircraft. The AD required airlines to add a secondary engine mount support, install a failure indicator for the bolt, and periodically inspect the indicator.

FAA becomes aware of most unsafe aircraft conditions through communications with aircraft manufacturers or reports of significant incidents or accidents. Using engineering judgment, FAA decides if these conditions warrant an AD. If needed, FAA engineers work with the aircraft manufacturer to develop the AD. The manufacturer usually prescribes the procedures needed to monitor and correct the unsafe condition, and FAA reviews and approves these procedures. FAA also establishes the time allowed for the airlines to comply with the AD. In determining the compliance time frames, which can range from immediate action to several years, FAA engineers consider factors such as the severity of the unsafe condition, the availability of parts needed to correct the condition, and the potential economic impact the AD will have on the aviation industry. In addition, some ADs require a one-time repair of an unsafe condition, while others require a combination of repetitive inspections, monitoring, and eventual repair.

FAA issues about 200 ADs each year for large commercial aircraft. At any one time, a particular aircraft type may have many ADs requiring recurrent inspections or repairs.

After issuing an AD, FAA monitors compliance through its airline inspection program. When FAA inspectors find AD noncompliance, the airlines must comply with the AD before operating the aircraft again. Inspectors may also recommend that FAA impose a civil penalty or take other administrative action against the airline.

FAA'S MANAGEMENT AND OVERSIGHT OF
AD COMPLIANCE NEEDS IMPROVEMENT

Recent accident investigations and special inspections by the NTSB and FAA found instances in which airlines have not complied with ADs. Because FAA's guidelines allow aircraft inspectors discretion in deciding when, what ADs, and how many aircraft to inspect for AD compliance, the inspectors do not always verify compliance during their routine inspections. In addition, FAA does not know the industry-wide extent of AD noncompliance because it does not receive information about the number of ADs inspected for or the number of airlines found in compliance.

Incidents of AD Noncompliance

Recent NTSB accident investigations and special inspections by FAA found that several airlines were not complying with ADs. For example, FAA issued an AD in October 1987 that addressed deterioration in the Boeing 737 fuselage, the aircraft type that was later involved in the Aloha accident. The AD required airlines to conduct visual inspections for cracking; and if cracks were found, the AD required airlines to perform additional technical inspections. According to NTSB's accident report, Aloha's records neither documented these required additional inspections nor accurately reflected the condition of the aircraft. Further, FAA records showed no evidence that its inspector had verified AD compliance. NTSB concluded that Aloha Airline's maintenance program and FAA's oversight of the airline's flight operations were inadequate. NTSB therefore recommended that FAA's inspection program place greater emphasis on evaluating the actual condition of each aircraft and the airlines' compliance with specific ADs.

FAA's own experience with direct inspection of airline operations has also found instances of serious AD noncompliance. In 1986, as a result of an Eastern Airlines aircraft accident, FAA conducted an in-depth review of the company's adherence to FAA regulations. This effort combined records reviews and direct aircraft inspections. FAA found, among other problems, that Eastern had flown two aircraft for almost five years without complying with an AD pertaining to the landing gear and without FAA inspectors' detecting the non-compliance. One aircraft made over 10,000 flights while not in AD compliance. The aircraft was damaged when the landing gear—which was the subject of the AD—failed. Another aircraft flew more than 8,900 flights while not in compliance with the same AD. Furthermore, during one six-day period, Eastern operated thirty-seven aircraft on over 1,100 flights without properly complying with an AD that required recurrent inspections for cracks. On the basis of this review, FAA imposed a $9.5 million fine for a variety of violations, including AD noncompliance. Also, FAA's 1987 "white glove" inspection of eight airlines found twenty-six instances of AD violations.

Monitoring AD Compliance

When issuing an AD, FAA relies on the airlines to comply with its requirements but does not require them to report compliance. Instead, FAA relies on its aircraft inspectors to monitor airlines' compliance during routine surveillance inspections. However, FAA's guidelines are too discretionary to ensure that critical safety areas, such as AD compliance, are adequately covered during inspections. We also found that FAA does not have information on the number Of ADs checked by inspectors or the extent to which airlines comply with ADs.

ADs are critical elements in maintaining aviation safety because they require airlines to correct potentially unsafe aircraft conditions. However, FAA cites AD compliance as one of more than twenty areas, such as pilot training and emergency equipment, that inspectors *may* consider including in an inspection. The guidelines do not specify how much emphasis inspectors should place on reviewing each inspection area, nor do they require inspectors to verify AD compliance during each inspection. In addition, the guidelines allow inspectors discretion to determine how often to conduct AD inspections, which ADs to check for compliance, and how many of the airline's aircraft to inspect.

Inspectors do not always verify AD compliance during routine inspections because some discretion in the inspection process is necessary to allocate FAA's limited work force among an increasing number of aircraft and hundreds of ADs. In addition, some types of inspections, such as en route and ramp inspections,[4] do not lend themselves to inspecting for AD compliance. Other types of inspections, however, such as maintenance records checks and spot checks of aircraft undergoing periodic service, are appropriate for verifying AD compliance. Because ADs prescribe corrective actions that airlines must take to correct known unsafe conditions in their aircraft and because several recent incidents suggest serious weaknesses regarding airline compliance with ADs, FAA needs to require its inspectors to test for AD compliance as part of each inspection.

We accompanied FAA inspectors on ten routine inspections. In only two instances did the inspector check for compliance with an AD during the inspection—once by reviewing maintenance records and once by comparing maintenance records to work done on the aircraft. In the other eight instances, due to FAA's discretionary inspection policy, the inspector decided to emphasize non-AD-related inspection duties, such as following up on reported mechanical problems during a flight, reviewing maintenance records for current work being performed on an aircraft, and reviewing training records for airline personnel, instead of checking for AD compliance.

FAA Needs More Complete AD Compliance Information. FAA does not know the extent of AD compliance throughout the airline industry because the agency does not have information on the extent to which inspectors check for, or find airlines in compliance with, ADs. Inspectors, in accordance with FAA's procedures, report instances in which they find that airlines are not in compliance; they are not, however, required to report when they observe AD compliance. Requiring inspectors to both verify AD compliance during inspections and report complete inspection results—compliance as well as noncompliance—would provide FAA with (1) a data base to measure the extent of AD noncompliance and (2) a management tool to help administer the program more effectively. Without such data, FAA cannot determine whether it has established appropriate emphasis and policies regarding AD compliance inspections.

FAA CAN MORE ACTIVELY USE AVAILABLE SAFETY DATA

FAA can enhance its management of the AD program by using available safety-related information to help focus its inspections and thereby improve the effectiveness of its limited inspector resources. Because FAA relies on inspectors to verify AD compliance, a more effective inspection effort would also help FAA identify where changes should be made in the AD program to further improve aircraft safety. The NTSB and the Airworthiness Assurance Task Force, a technical panel of industry and government aviation experts, both recently concluded that FAA's inspection efforts are largely reviews of paper records with limited hands-on aircraft observation. FAA acknowledges it would prefer more hands-on inspections, but it cites manpower limitations as precluding it from inspecting all aircraft and verifying compliance with all ADs. Therefore, FAA must judiciously allocate its limited resources to achieve the maximum effectiveness from its inspection program.

Analyzing existing safety-related data, such as in FAA's Service Difficulty Report system, could provide this focus by raising warning signals regarding specific aircraft, aircraft types, or airlines that warrant a closer hands-on inspection. FAA uses these data to help focus its special inspections but does not as a matter of course use the data for routine airline surveillance.

FAA has maintained extensive safety information in various data bases. For example, FAA requires airlines to report mechanical problems that occur in aircraft, such as problems with landing gear, identification of corrosion and cracking, and engine shutdowns, to its Service Difficulty Report system, which has existed for over two decades. In 1988, commercial airlines reported approximately 19,000 mechanical difficulties to the system.

We analyzed Service Difficulty Report data by aircraft type, such as the Boeing 727 and McDonnell Douglas DC-9, to determine whether this information could be used to augment FAA'S inspection process. We reviewed a limited sample of reports that airlines had submitted to FAA between January 1983 and June 1989 and noted a wide range in the frequency of problems reported across aircraft type and airlines. For example, in reviewing Service Difficulty Report data for 727s, we noted that one airline reported thirty-seven instances of structural problems involving cracking and corrosion found during inspections of one of its aircraft. All of these problems were found during eight aircraft inspections that began in September 1984 and concluded in January 1988. During the last inspection, the airline found and repaired twenty-eight instances of cracks and corrosion in areas such as floor beams, brackets, and fuselage skin. However, the airline did not submit any additional Service Difficulty Reports to FAA from January 1988, when repairs were made, to June 1989, the ending date of our analysis. If FAA analyzed these safety data, the agency could better focus its inspection efforts by selecting for reinspection this or other aircraft with similar findings to determine whether any additional cracking and corrosion have occurred since the last reported inspection.

FAA could also use Service Difficulty Report information to monitor whether airlines are finding and correcting generic problems, such as cracking and corrosion, that can occur in aircraft. For example, for one type aircraft, we examined six Service Difficulty Reports that airlines had submitted to FAA. Five of the reports stated that the airlines had found and corrected cracks and corrosion in their aircraft. Airlines operating other aircraft makes and models also reported finding these and other types of deterioration.

While this type information cannot be used exclusively to determine problems with specific aircraft, aircraft type, or airlines, FAA can use the data as a warning signal of potential safety problems. FAA inspectors could use these signals, along with their knowledge of other factors—such as aircraft age, utilization, airline maintenance practices, and various types of normally expected deterioration—to determine which specific aircraft, aircraft type, or airline should receive increased inspection emphasis. FAA notes that its reporting system may have some problems regarding the quality and timeliness of the data. In a separate review, we are currently assessing FAA's Service Difficulty Report system to determine how to make these data more useful.

In 1987, we issued two reports that identified improvements needed in FAA's airline inspection program. In one report, we recommended that FAA use safety data to target its inspector resources toward high-risk conditions.[5] In the other, we recommended that FAA revise its inspection guidance to take into account the need to target airlines displaying characteristics that may indicate safety deficiencies.[6] Consistent with these recommendations, FAA could use the kind of data maintained in the Service Difficulty Report system, as well as other systems and data bases maintained by the agency, to develop indicators of potentially unsafe conditions in aircraft as tools to help target its limited inspection resources.

CONCLUSIONS

Because by their nature ADs address critical safety conditions, airline compliance with AD requirements and FAA's ability to effectively oversee and manage the AD program are vital to aviation safety. Poor implementation of AD requirements can potentially result in aircraft accidents. This is illustrated by the Eastern Airlines incident involving the failed landing gear. Furthermore, the NTSB investigation of the Aloha accident and FAA's special inspection results demonstrate that (1) some airlines are not complying with ADs and (2) AD noncompliance can remain undetected by FAA for long periods of time.

FAA does not compile information on the extent of AD noncompliance because (1) the number of ADs inspected for is left to the discretion of each aviation safety inspector and (2) the inspectors report only when airlines are not in compliance. Consequently, FAA is not aware of the number of ADs inspected for or the number verified as being in compliance. This information is needed, however, to measure the extent of AD compliance throughout the airline industry. The instances of noncompliance discovered by recent accident investigations and special FAA inspections, coupled with the wide discretion given inspectors in verifying ADs and incomplete inspection reporting, indicate a need for FAA to improve its management and oversight of this critical safety program. FAA needs to know if its inspectors are checking for AD compliance and the extent to which airlines are complying with ADs. With more complete information, FAA would have the management tools needed to determine whether (1) airlines are complying with ADs and (2) additional regulatory action is needed to ensure AD compliance.

Also, FAA can improve the management of its AD process by analyzing existing safety data to help determine where to focus its limited inspection resources. FAA can use this information to identify which aircraft, aircraft type, or airlines warrant additional hands-on inspection emphasis.

RECOMMENDATIONS

To improve FAA's management and oversight of the AD program, we recommend that the Secretary of Transportation direct the administrator, FAA, to

- require a systematic testing for AD compliance as part of each routine airline inspection,
- require inspectors to report which ADs are tested and the extent of airline compliance found during each inspection, and
- maintain and analyze compliance information to determine the extent of AD noncompliance and any additional actions necessary to ensure that airlines comply with ADs.

We also recommend that the Secretary direct the administrator, FAA, to analyze and use available aircraft safety data as a management tool to focus FAA's limited inspection work force.

We discussed the facts presented in this report with cognizant FAA officials. They agreed that the report is accurate. However, as requested by your office, we did not obtain official agency comments. Details on our objectives, scope, and methodology are contained in appendix I [not included in this volume].

NOTES

1. *Aviation Safety: FAA Has Improved Its Removal Procedures for Pilot Examiners* (GAO/RCED-89-199, Sept. 8, 1989).

2. *Aviation Safety: FAA's Safety Inspection Management System Lacks Adequate Oversight* (GAO/RCED-90-36, Nov. 13, 1989).

3. *Aviation Safety: FAA Is Considering Changes to Aviation Medical Standards* (GAO/RCED-90-68FS, Jan. 9, 1990).

4. An "en route" inspection is a check of an aircraft during a flight and includes observing the flight crew's usage of aircraft equipment and the performance of the equipment. A "ramp" inspection is a check of an in-service aircraft and includes observing the refueling of the aircraft, passenger handling, and the condition of the aircraft.

5. *Department of Transportation: Enhancing Policy and Program Effectiveness Through Improved Management* (GAO/RCED-87-3, Apr. 13, 1987).

6. *Aviation Safety: Needed Improvements in FAA's Airline Inspection Program Are Underway* (GAO/RCED-87-62, May 19, 1987).

22

The FAA, the Carriers, and Safety*

The topic in this chapter concerns the technological forces that drive the air transport system. I offer the following hypothesis most tentatively, because I am not certain that the many clear exceptions to it do not overwhelm the supporting cases, but it is a hypothesis worth consideration. The hypothesis is that the air transport industry (aircraft manufacturers and the airlines) supports safety regulations and requirements primarily when the increase in safety permits an increase in production efficiencies, and that the FAA concurs in this strategy. The industry is not against safety, and does a lot to increase it on its own; it is, after all, a prerequisite of the system that it be reasonably safe. But it will voluntarily undertake safety modifications primarily under two conditions: (1) when the modifications make increases in production efficiency possible (building more economical aircraft and engines, for the equipment side of the industry,

---

* Reprinted with editorial changes from *Normal Accidents: Living with High Risk Technology*, by Charles Perrow by permission of Basic Books, Inc., Publishers, New York. Copyright © 1984 by Basic Books, Inc.

and increasing density and decreasing operating costs, for the service side) and (2) when they can be added to new aircraft without significant cost, especially if there is fear that a retrofit of the equipment might be required by public pressure (largely through Congress) or (more remotely, as we shall see) by FAA requirements. This means that voluntary safety modifications or additions will not be made simply because there is evidence they are needed. The industry will concur in and not protest and delay *mandatory* safety efforts primarily when these increase efficiency of the system (including higher utilization by the public).

All that this careful wording really says is that no one in the industry is going very far out of their way to protect the lives of employees and customers and innocent bystanders (first-, second-, and third-party victims). Perhaps it will always be thus, and we should not be surprised to find this attitude in an activity that is primarily for-profit in nature, and furthermore must be organized through large, formal organizations (which inescapably will be indifferent to some degree to the fate of these victims). Yet the rhetoric of the industry and the FAA sharply contrasts with this view, and thus, the hypothesis needs exploring and airing. This will do this, though after perusing it the reader may board her next commercial flight with less than her customary ease.

Safety involves two factors—accident prevention, and damage mitigation after an accident. The industry and the FAA have been preoccupied with the former, because each improvement there has meant greater density, higher speeds, and more customers. The latter, damage mitigation, has little or no effect upon these economic variables. It merely reduces the injury and death rate from accidents. The largest injury and death rate sources in damage mitigation come from evacuation delays, and more important still, from cabin missiles, obstructions, and toxic fumes and explosions.

First, let's examine the matter of timely evacuation of aircraft after an accident. A key to this task is a functioning public address system and means of communicating with the flight attendants. American passengers are remarkably compliant when faced with uncertainty in these awesome technological marvels. They will sit still until told to get out. If the electrical system is damaged, or all power shuts off to reduce explosion dangers, or the craft runs out of fuel, there *is* no way to tell the attendants or the passengers to get out. In 1971 a Boeing 747 caught fire after an aborted takeoff and landed again. The first officer made an evacuation announcement, but inadvertently made it over the radio rather than the public address system. When nothing happened in the passenger cabin, he tried the proper system, but it was inoperative because all power had been turned off to reduce the risk of fire and explosion. The flight crew then entered the passenger cabin and shouted the order, but only those passengers in the front part of the cabin heard. All eventually evacuated safely, but since this had happened before, the NTSB recommended to the FAA that self-powered audio and visual alarm systems be installed. The FAA agreed it was needed, but believed further study should be made of the best system to require.[1] This report was presented in 1972; and the FAA continued its "study."

After some more accidents where the public address system failed and passengers were needlessly injured or killed, the NTSB tried again to get the attention of the FAA through a special study of their own, with recommendations, in 1974. Nothing happened. The Safety Board reiterated its recommendations after the 1975 crash of a DC-8 in Portland, Oregon, where the system was inoperative. Six years later, on January 19, 1981, the FAA finally issued a proposal for requiring self-powered warning systems, but still had it under review at the end of 1981. Air carriers would have a full two years in which to comply, thus providing a minimum of thirteen years for the studies, recommendations, and implementation for a system as simple as a battery-powered speaker system. The cost of the equipment is estimated to run from $500 to $5,000 per aircraft. Some aircraft already have the system in operation—United had it on four of its five aircraft types, but other airlines did not. Even the regular public address system does not have to be repaired before twenty-five hours of flight are up for some aircraft, and for others, such as McDonnell Douglas aircraft, there is no limit to the time that the airline can take to repair a malfunctioning address system. The NTSB deplored these lax regulations, to no effect.

More serious is the matter of cabin safety.[2] The predecessor agency of the NTSB, the Civil Aeronautics Board, recommended to the FAA in 1962 that the testing that was going on in the FAA regarding seat failures during a crash be expedited. This followed a 1962 crash where it was believed that twenty-eight persons could have been saved if their seats had not ripped out. The FAA replied that it recognized the need for further studies, and it was pursuing them "consistent with available manpower and funds." (Stronger bolts could have made a large difference, and extensive studies to determine that were hardly needed.) The regulations regarding the crash force that seats would have to withstand were then ten years old, and aircraft had become larger, faster, and were crashing with more force. By the end of 1981, when the NTSB undertook another special study of the problem, the old 1952 standards were still in effect.

The special study found that since 1970, only examining those crashes where all or at least some passengers could be expected to have survived the force of the impact itself, sixty percent of the crashes exhibited failures of cabin furnishings. Of the more than 4,800 passengers involved in these crashes, over 1,850 were injured or killed. Many of these deaths and injuries could have been prevented, the study concluded, had cabin furnishings not failed, particularly in the 46 percent of these accidents where there was fire.

Of the forty-six accidents where cabin furnishings failed, seats or the seat belts failed in 84 percent of the accidents; overhead panels and racks failed in 77 percent; galley equipment in sixty-two percent. Most of these failures occurred when the g (gravity) forces were well below the figure that the FAA sets as the maximum survivable force, and under which the equipment should survive. (In the John Wayne Orange County aircraft accident we described earlier, there were four serious and twenty-nine minor injuries caused by seat failures or other cabin furnishings, though the g forces were well below the standard of the 1952 regulations.) However, the study conclusively showed that the FAA maximum was far too low; people survived much higher g forces than the forces the FAA set as the maximum for survivability, and thus the seats and other furnishings should also be required to survive these higher forces. While this fact had been well established for some years, even by the FAA's own studies, it was still disputed by the FAA in congressional testimony in 1980. The FAA currently has a huge study underway, started in 1980 and not expected to be concluded until 1985. (It might recommend stronger bolts.) The NTSB comments: "Although it should be possible to conduct many worthwhile experiments and to gather new data in this test, the Safety Board questions whether the FAA will be any more willing to accept such crash data as being representative of modern aircraft." It also argues that "the major emphasis of FAA's ongoing crashworthiness programs should be on applying available technology. . . ."[3]

The problem is not only flying missiles, flying seats with occupants, jumbled debris preventing evacuation, and inoperative exits. Toxic fumes are probably the major killer. When a Saudi Arabian flight exploded and burned on the runway at Riyadh in 1980, killing the crew and 301 passengers, it was the smoke and toxic fumes created by cabin furnishings that proved lethal. Cabin materials, when heated or burned, produce deadly hydrogen cyanide and hydrogen chloride, which produces hydrochloric acid, phosgene (a nerve gas when ingested), and an explosive high-temperature mixture which consumes all oxygen and leaves only carbon monoxide.[4] At least 371 persons in recent years are known to have survived crashes only to die as a result of fires involving cabin material. The first such fire occurred in 1961, but the FAA has been reluctant to require the use of flame-resistant materials. The chairman of the NTSB, James King, said in 1980, "Ever since the 1961 crash . . . the FAA has promised action. No action has been forthcoming."[5] The National Research Council of the National Academy of Sciences reported in 1977 that safer resins and foams were available for use. Even such painless improvements as the elimination of carpet as a wall decoration would help, it said.

The FAA has overlooked short-term improvements in the search of an elusive, perfect solution, notes Jeffrey Smith in an article in *Science.*[6] More important, the General Accounting Office of Congress notes, the FAA has issued proposals twice, but withdrawn them because the industry was opposed. The FAA convened a panel of 150 of the world's top experts in aircraft fire safety, but about 100 of them were from the industry itself, and the FAA. After two years it concluded that the FAA was on the right track in this area. With this backing the FAA proceeded cautiously, contracting with one of the aircraft manufacturers to develop a highly sophisticated fire chamber at the cost of about a half a million dollars. Even this use of public funds was not enough; the FAA decided it was not sufficiently sophisticated and that more money and at least another year were needed. Meanwhile, it continues its own testing, which involves holding a Bunsen burner (1952 model, no doubt) to cabin material to see if it burns. The problem is extreme heat, which decomposes the material, not the presence of a cigarette lighter. A radiant heat panel test has long been advocated by the National Academy of Sciences and other groups.

What is going on here, in an agency that developed and installed the sophisticated air traffic control system and is launching an even more automated and advanced one? The conflict between the NTSB and the FAA is, perhaps, to be expected, since the NTSB is the independent agency set up to review accident reports, conduct background studies, and recommend to the appropriate federal agency (the FAA in the case of air transport) changes in regulations, more intensive research, and so on. The forerunner of the NTSB was the Safety Bureau of the Civil Aeronautics Board, but it was made independent of the regulatory agency when the National Transportation Act was passed in the mid 1960s, so that the same person (in this case, Jerome Lederer) was not both promulgating civil air regulations and investigating the accidents they might cause. The FAA replaced the CAB, but over the years it has been criticized as being too close to industry. The General Accounting Office of Congress, a House government operations committee, the Ralph Nader-affiliated Aviation Consumer Action Project, and other groups have recently charged that not only is the FAA too industry oriented, but the Reagan administration has cut back on the funding of its primary critic and watchdog, the NTSB, and the FAA has relaxed many rules and restrictions (for example, commuter airline pilots can now work a seventy-hour week; pilots for the large airlines are restricted to thirty). The air transport industry, through its various trade associations, vigorously supports the policies of the FAA. But why would the industry, and the FAA, drag their feet on cabin safety, if the critics are correct in their charges? It appears that the air transport industry welcomes and supports efforts to allow more efficient, economical, and reliable flights, and these efforts improve safety. But bolting seats down better, or using inflammable material for decorations will not increase efficiency, nor is it likely to increase ticket sales. The airline cannot even be sued for negligence in these respects. (A jet airliner may fly fifteen to twenty flights with an inoperative public address system without liability or penalty.) It is not that the proposed improvements would cost much. Industry is sometimes willing to put them into new planes, and even retrofitting costs are not high. It just seems as if they are either a nuisance, or that tough regulations would establish a precedent for the FAA that the industry fears. How else can we explain a two-year study by a group dominated by industry and FAA representatives that concluded the FAA was doing a fine job in not upgrading thirty-year-old standards that resulted in perhaps hundreds of needless deaths?

NOTES

1. NTSB, SIR-81-6, 9 September 1981.

2. Ibid.

3. Ibid., pp. 30, 32.

4. Jeffrey R. Smith, "FAA Is Cool to Cabin Safety Improvements," *Science* (February 6, 1981) p. 557.

5. Ibid.

6. Ibid., p. 558.

23

International Airline Passengers Association Critique of the DC-10*

November 27, 1989

The Honorable Tim Valentine, Jr.
1510 Longworth Building
Independence and New Jersey Avenues, S.E.
Washington, D.C. 20515

Dear Mr. Valentine:

These comments are submitted pursuant to hearings conducted by the Transportation, Aviation and Materials Subcommittee on November 9, 1989. Although IAPA was not asked to testify, we have been closely associated with the subject areas addressed by the sub-

---

* Reprinted by permission of the International Airline Passengers Association.

committee. Specifically, these hearings focused on engine failures associated with the DC-10 aircraft, and served to explore specific steps that might be taken in order to preclude any recurrence of the type accident that occurred at Sioux City, Iowa on July 19 of this year. Dr. Jim Green, of the subcommittee staff, has asked that we submit written testimony for the record in order that our views and recommendations may be considered in the course of your deliberations. Accordingly, the following comments and recommendations are submitted for the subcommittee's consideration.

We are well aware of the excellent safety record compiled by the world's commercial carriers, and we are not prone to over-reaction or alarmist measures when an industry problem is identified. Rather, in the interest of our members and airline passengers at large, we seek to help resolve those matters that concern the flying public. When safety of flight becomes an area of concern, our commitment knows no bounds and, along with others in the industry, we are determined to protect the public's confidence in commercial aviation.

A recently completed survey of our membership indicates a genuine and deepening concern about the DC-10 aircraft. Obviously, much of this concern has been generated by the Sioux City tragedy and a number of operational incidents that occurred during that same general time frame. Along with the rest of the industry, we await the results of the NTSB's investigation, and findings by FAA and other government examiners. Usually, these results are published twelve to eighteen months after the occurrence. Meanwhile, FAA has established a task force to examine all wide-body aircraft, including the B-747 and L-1011 aircraft. The B-747 is included because of the JAL tragedy, even though we understand Boeing, of its own initiative, made safety-related changes to the aircraft. The L-1011 is included because of past non-fatal engine containment problems. Unfortunately, this broad, sweeping scrutiny of the wide body aircraft fleet has caused our membership to question the airworthiness of transport category aircraft in general.

Consumer concern about the DC-10 aircraft, in particular, is understandable when we consider past incidents involving this aircraft:

- In 1974, a Turkish Airlines DC-10 crashed near Paris after experiencing explosive decompression. The aircraft floor buckled and severed strategic control apparatus, rendering the aircraft uncontrollable with the resultant loss of 346 lives. The FAA failed to pursue timely corrective action following that tragedy.

- On May 25, 1979, an American Airlines DC-10 crashed on takeoff at Chicago's O'Hare Field following separation of the number one engine from the wing, rupturing hydraulic lines that deactivated the left wing slats. The aircraft was uncontrollable and 273 people died.

Following the Chicago crash in 1979, IAPA (then known as the Airline Passengers Association) petitioned the FAA to order a "thorough scientific inspection" of the entire engine support system of the DC-10 fleet. The FAA failed to respond to that request and to subsequent requests for information relevant to that tragedy. On June 3, after repeated unanswered requests for immediate and decisive corrective action, IAPA filed a motion in federal court to ground the DC-10 fleet. That motion was quickly granted, and the fleet was grounded for more than one month while nonscientific inspections were completed on selected aircraft.

Some corrective measures were subsequently ordered by the FAA, but one of the carefully researched recommendations IAPA offered to the FAA, unfortunately, went unheeded. Specifically, IAPA wrote:

> The various parallel hydraulic and electrical systems of the DC-10 should be sufficiently separated from each other, either by structure or distance, so that in the event of accidental damage to one system, it would be unlikely that the other backup systems could be affected by the same source of damage.

- Then, on July 19, 1989, United Airlines Flight 232 experienced total hydraulic failure following the uncontained failure of number two engine, with the resultant crash at Sioux City claiming 111 lives.
- Since the DC-10 entered the commercial fleet in 1971, there have been multiple accidents and incidents, some with catastrophic consequences. Obviously, these events can be attributed to many factors. However, it is worth noting that at least seventeen of the 445 DC-10s delivered—or 3.8 percent of the fleet—have been wrecked. For purposes of comparison, this contrasts with 1.2 percent of the L-1011 fleet, and 1.5 percent of the B-747 fleet.
- Since the Sioux City tragedy, another DC-10 has crashed in Libya with , a loss of at least seventy-seven people; a United DC-10 with 240 people on board experienced hydraulic failure but landed safely at Los Angeles; a Canadian Airlines DC-10 carrying 254

people lost a main landing gear wheel, but landed safely at Toronto; an American Airlines DC-10 had to be towed to the gate at Dallas-Fort Worth airport due to a loss of hydraulic pressure; and American Flight 621, forty-five minutes into its flight to Las Vegas, returned to Dallas-Fort Worth Airport because of low oil indications and subsequent engine shutdown.

We do not contend—or even suggest—that a common cause exists for the multitude of accidents and incidents that have plagued this aircraft since 1971. Moreover, IAPA is not a newcomer to the aviation safety business, so we fully appreciate the sacred aspects of ongoing accident and incident investigations. We have not, nor will we ever, be a party to premature speculation about the cause of aviation accidents. On the other hand, we are acutely aware of the DC-10's past operational difficulties and the continuing sequence of recurring fleet problems. Frankly, we are deeply concerned, even fearful, that lives may be imperiled at this very moment. I'm convinced that decisive and immediate action is imperative, and any delay is not in the best interest of either the FAA or the traveling public. We believe that many years ago the FAA was offered a rare opportunity to fulfill its aviation safety responsibility as guardian of the public trust. Regrettably, the challenge went unanswered and the FAA elected to take the path of least resistance. Now, in light of the latest DC-10 failure, we are once again petitioning for aggressive action.

We have suggested that the DC-10 fleet must be grounded if that is the only way to identify and correct existing deficiencies, while protecting the flying public. I can assure you that we have no interest in disrupting air commerce, but recent events have done nothing to heighten our confidence in this aircraft, or in any way alter our views about the DC-10 problem. On the other hand, if there are measures, short of grounding the fleet, that would resolve these problems while protecting public safety, IAPA and its 110,000 members would certainly be supportive of those efforts.

The DC-10 is unique among American-produced wide-body aircraft in that it is subject to total loss of control with severance of hydraulic lines which are used to control many primary flight surfaces. The Sioux City accident brings to a total 730 people killed in three crashes of the DC-10 when events resulted in lack of control of the aircraft. The lack of controllability was demonstrated by flight simulator tests which was testified to in Sioux City by a McDonnell-Douglas Engineering test pilot. He stated that loss of all hydraulics "is not a trainable maneuver" and that the aircraft goes into oscillations of about 50 seconds, like a roller coaster. No volunteer pilots were able to land the aircraft in simulator flights with loss of controls.

Check valves are now being suggested for installation in one out of the three hydraulic systems on the DC-10 aircraft. McDonnell Douglas proposes to install an electrically operated check valve in the number three hydraulic system actuated when the hydraulic fluid reservoir reaches a critical level. The actuation of the valve would be displayed by an enunciator light on the flight engineer's panel. The FAA has granted nearly two years in order to modify the fleet of 450 DC-10s to protect this system. Plans are to have a prototype in February, 1990, with production of about fifty units a month to reach May, 1990, and one hundred per month thereafter for finishing up the fleet on August 15, 1990, with installation in the following year. We contend that this is simply too slow. Clearly, time is a vital factor and action must be taken quickly. While we recognize and appreciate the value of a joint government/industry task force to review design of wide-body jets, this is hardly the appropriate forum to address the urgent and critical aspects of the DC-10 problem. Nor do we believe it is reasonable to permit nearly two years to elapse before accomplishing a basic modification to the number three hydraulic system of the DC-10 fleet. In our judgment, the following additional actions should be undertaken without delay, in order to insure that the interests of the flying public are protected:

- Immediately shield the horizontal stabilizer critical damage area with hydraulic hose made of *Kevlar* or other similar material of proven strength and durability.
- Proceed with installation of modifications recommended by McDonnell Douglas for the DC-10, and make the modification applicable to the MD-11 as well. Complete the program within twelve months.
- Require re-installation of engine vibration monitor instruments which were removed from DC-10s and other aircraft. Correct the occasional false alarms experienced with this system.

Finally, we believe the following questions are pertinent to the DC-10 issue.

- In 1977, UTA airlines experienced an uncontained engine failure leading to inadvertent slat retraction on the DC-10. Why was the danger of similar accidents occurring not taken into account, and necessary changes ordered before the Chicago disaster of May 25, 1979, and the near disaster of Air Florida's DC-10 in September, 1981?

- In 1982, after the Air Florida DC-10 near disaster at Miami in September of 1981, the FAA undertook responsibility for establishing new standards for containment of engine discs and rotor blades so that structural damage would not be caused to the aircraft. Why, then, did it take until 1988, some six years, before Advisory Circular 20-128 was issued for methodologies to minimize a hazard of damage from uncontained engine failures?

- The National Transportation Safety Board has issued Class Four Urgent Category safety recommendations A-89-95 through 97 following the Sioux City crash, regarding inspection of engines on an urgent basis. Why has the NTSB not issued safety recommendations with regard to installation of check valves or other protective measures for the hydraulic systems on the DC-10, on an equally urgent basis?

- In light of the Lockheed modifications of check valves on the L-1011 wide-body aircraft in 1972, and the modification of the Boeing 747 hydraulic systems by "hydraulic fuse" valves after the Japan Airlines crash in 1985, why did the FAA and NTSB not make recommendations earlier to the effect that other wide-body manufacturers install such systems?

- What has been the experience factor with the use of *Kevlar* material for containment rings, with regard to effectiveness and weight?

- Is it true that the cost of the modification made following the Air Florida uncommanded slat retraction involved approximately $10,000 parts and labor per DC-10?

- Are the hydraulic check valves required by the 1981 airworthiness directive for DC-10 slat modification the same type that are being suggested for the number three system?

- Has *Kevlar* been tested for use as an alternate to the current hydraulic hose material? If so, what were the findings?

- Is it true that the DC-10 involved in the UAL Flight 232 crash was equipped with engine vibration monitor for each engine at the time of delivery and those were subsequently removed?

When IAPA demanded grounding of the DC-10 fleet in 1979, our sole objectives were identical to our present goals—first, find out what's wrong with the aircraft; and secondly, get it fixed. IAPA's only interest in this matter is safety of the flying public, and I'm confident that we have your support in our pursuit of this common goal. Unfortunately, delay and indecision in the past have denied FAA the opportunity to clearly demonstrate its firm commitment to consumer safety and protection of the air traveler. For this reason we solicit your leadership in insisting on immediate and effective corrective action for the DC-10 aircraft.

The International Airline Passengers Association (IAPA) is a membership organization comprised of more than 110,000 frequent business flyers, representing over 175 countries around the world. Our members are professional people who regularly and almost exclusively utilize air transportation for business and leisure travel. They collectively purchase over four million airline tickets and average over forty round trip flights per year on commercial aircraft. They are a significant part of the eleven percent of air travelers who account for almost half of annual airline revenues.

Thank you for the opportunity to submit these comments for the public record. I know you share our desire to preserve public confidence in commercial air transportation, and your efforts have helped in attainment of that goal.

Sincerely,

Richard E. Livingston
Director of Operations

24

Moral Responsibility for Engineers*

I

Technological knowledge gives us new and greater powers both to benefit and, as we have become increasingly aware, to harm ourselves. How are we to obtain the benefits and avoid the harm? One popular answer focuses on the individuals who design the products, conduct quality and safety tests, and directly oversee production. These individuals—practicing engineers, technicians, low-level managers[1]—it is claimed, have to be guardians of society. They are thought to stand under a special—if not exclusive—moral obligation to protect society from the harms that could result from technological development. They must be ready and willing to risk their jobs and make other personal sacrifices in order to protect and promote public welfare.

This conception of engineers' moral responsibility is common

* Reprinted with editorial changes by permission of the author. Copyright © 1983 by Kenneth D. Alpern.

not only in the public mind but among engineers themselves—at least in their public pronouncements. (Virtually every one of the many codes of ethics proposed by engineers gives paramount place to serving the public good.[2]) But does this conception of the moral responsibility of practicing engineers survive criticism? Given the nature of engineering activity and the nature of the organizations in which it is normally carried out, aren't these demands excessive, misplaced, and unrealistic? Why should engineers have to bear a *special* burden for the benefit of society? Indeed, can't it be argued that the nature of the corporate organization,[3] with its specialization of labor and delegation of authority, insulates practicing engineers from moral responsibility? And, even if there is an ideal according to which engineers are guardians of society, isn't the suggested moral demand in fact unreasonable and impractical given the realities of corporate competition and job insecurity?

In this essay I shall defend a strong conception of the moral responsibility of practicing engineers and the others I have mentioned who are active in the creation and production of technology. I will argue that though engineers are bound by no special moral obligations,[4] ordinary moral principles as they apply in the engineer's circumstances stipulate that they nonetheless be ready to make greater personal sacrifices than can normally be demanded of other individuals. Having made that argument I will defend it against common objections that seek to show that this demand is misplaced or unreasonable.

The "ordinary moral principles" stated in section II below are, I claim, quite general in scope, applying to anyone whose actions may contribute to harming others. However, to argue adequately for that claim would require a much longer essay. Instead, a few examples are provided as suggestive. The objections to the application of the principles, which I turn to in section III, are those most often heard in defense of or as excuses for questionable actions by engineers in corporations. These objections and the responses to them can, I think, easily be extended to cover the others mentioned above who are directly involved in developing and producing technological artifacts. I suspect, though, that further, often involved, arguments would be necessary to defend the principles in application to others in corporations (e.g., lawyers, accountants, janitors, secretaries), whose work is less directly related to technological production, and to professionals who are not in the corporation but hired by it for special purposes.[5]

Before turning to the arguments, I want to make perfectly clear just what sort of thesis I am arguing for. My thesis is that engineers have a strong moral obligation, where strength of obligation is to be understood in terms of the degree of personal sacrifice that can be demanded. I will say nothing, however, about how the public welfare is to be determined and little about the sorts of circumstances in which moral responsibility is to be exercised. These are quite different tasks from the one I have set and they require their own careful treatment.[6] My main aim is just to *establish* that engineers must meet a high standard of moral performance. Detailed argument on the proper exercise of their moral responsibility awaits another occasion.

II

I begin from the assumption that, other things being equal, it is morally wrong to harm others. This principle is a feature of popular morality (e.g., it is a sentiment behind the Golden Rule) and of virtually all types of ethical theory—utilitarian, intuitionistic, contractarian, axiological, etc.—and I take it that no argument for it is necessary here.[7] Somewhat more carefully stated, the principle—which I will call the Principle of Care—is this:

Other things being equal, one should exercise due care to avoid contributing to significantly harming others.

This principle has, of course, only *prima facie* force and is acceptable only with certain qualifications that would allow for legitimate competition, the voluntary assumption of risks, etc.

The "due care" clause of the principle designates such things as apprising oneself of the harm that may result from one's actions, taking precautions to avoid such harm, and being ready and will to make sacrifices in order to reduce the likelihood of harm. One may think in terms of the responsibilities assumed in driving a car: one must recognize the dangers, attend to the driving, be skillful in controlling the car, and be willing to risk one's property and, to some extent, person in order to avoid injury to others.

The Principle of Care is stated in terms of *contributing* to harm rather than in terms of *causing* harm because, as I see it, we have a basic obligation to avoid playing *any part* in the production of harm: both to avoid playing a direct causal role and to avoid creating conditions from which harm can reasonably be expected to arise.[8] Either sort of failure can result in culpability.[9] Of particular importance in the present context is that moral obligation does not hold only of those who "make the decisions." More on this point later.

As the Principle of Care stands, it is vague. What counts as due care? When is harm significant? What are the relevant types of contribution to harm? A bit of the vagueness of the principle can be removed by noting that the degree of care due is a function of the magnitude of the harm threatened and of the centrality of one's role in the production of that harm. For example, the driver of a gasoline truck must be more attentive and have greater skill than an ordinary motorist. This is true for the simple reason that the truck driver is in a position to create greater harm than is the ordinary motorist. Similarly, journalists must exercise special care in their behavior because of the critical role they may play in the formation of the beliefs and attitudes of the public. The general rule here, a corollary to the Principle of Care, may be called the Corollary of Proportionate Care:

> When one is in a position to contribute to greater harm or when one is in a position to play a more critical part in producing harm than is another person, one must exercise greater care to avoid so doing.

The consequences of the Principle of Care and the Corollary of Proportionate Care are, I think, easy to draw. Practicing engineers exercise considerable control over technological developments. Though they may not be the ones who decide which projects shall be worked on, they do design, test for quality and safety of, employ, and maintain technology. Their actions (or refusal to act) can greatly affect public welfare. And, given the nature of much technology, their work affects the public welfare, for better or worse, to a greater extent than do the activities of most other citizens. Managers and higher-level executives, of course, likewise affect the public welfare (through directing rather than creating and testing), and so they are similarly subject to this higher standard. Thus, by the Corollary of Proportionate Care, practicing engineers can be held to a higher standard of care; that is, it can be demanded that they be willing to make greater sacrifices than others for the sake of public welfare.

This higher standard against which engineers are to be judged does not require supererogation of them; it is merely the consequence of the *ordinary* moral requirements of care and proportionate care as they apply to the circumstances of engineers. However, since significant disincentives and obstacles will often stand in the way of meeting these ordinary moral requirements, engineers can expect to have to exhibit a certain degree of moral courage in the course of their everyday work. This does not mean that any and all demands on engineers to sacrifice their personal good for the public good are justified. But it does raise the standard for them.

These principles do not single out engineers. The same principles and conclusion hold for anyone in a position of power—power to harm—from truck driver to president. If one is not willing and able to make the sacrifices, then one should not seek or accept the position of power. To do so would be to act immorally.

III

Many objections can be raised against the application I have made of the Principle of Care and the Corollary of Proportionate Care to the circumstances of practicing engineers. For the rest of this essay I will consider several attempts to deflate engineers' responsibility or to shield them from it.

"I'd lose my job if I didn't."

Criticized for acting immorally at his job—say, for falsifying a report on safety tests—an engineer might offer the excuse, "But I'd lose my job if I didn't. It's unfair to expect me to jeopardize my job." This response is just a version of the immemorial complaint that morality often conflicts with self-interest. The traditional reply is to point out that moral considerations are just the sort of things that override self-interest and that is that: it is time to exhibit moral courage.

Though I am arguing in this essay that engineers must be prepared to exhibit moral courage, I think the situation is a bit more complex than the traditional response allows. Two sorts of cases must be distinguished. First is the case in which the engineer himself conceives the idea of falsifying the report in order to gain advantage for himself. It is to this sort of case that the traditional response most directly applies, and that, indeed, is all that needs to be said. The pursuit of personal gain (greed) is no excuse for wrongdoing.

But now consider the sort of case in which the initial impetus for falsifying the report comes from another person who has authority over the engineer. Taking the crudest case (from which the main point emerges most clearly), we can imagine the engineer's supervisor to have threatened to fire him if he doesn't sign the report. It is still wrong, and a breach of moral integrity, for the engineer to sign the false report, but, in the circumstances, he deserves sympathy, for he is in one sense a *victim* faced with having to make a sacrifice because of *another* person's immorality.

Now the question is, Should the engineer receive more than sympathy? Should he receive positive support of some kind from society? Why, the engineer may ask, must he alone and unsupported have to bear the burden of vigilance and sacrifice for society's sake? One answer is that he has entered and continues his employment voluntarily, full well knowing that his chosen job has certain benefits and certain liabilities. One of the liabilities is that his moral integrity will likely be put to the test. Recognizing this, he is free to choose not to take that particular job or not to embark on that sort of career. But if he does choose that employment, then the responsibility to bear the burden is solely his.

This answer is, I think, basically sound, but qualifications are necessary. For one thing society *needs* the moral vigilance of individual practicing engineers. Government regulation, for example, does not replace it: regulations cannot be sufficiently detailed, flexible, or up to date; regulators who enforce the rules cannot be present to evaluate each significant decision. (And besides, what would insure that the regulators would apply the regulations conscientiously?) So, society needs moral engineers. How are moral engineers to be secured? Higher salaries may attract some, but, by and large, if it's moral engineers that we want, then we will have to make it possible to practice engineering morally, and this means providing support for engineers when moral action is difficult.

There is a further argument not only that society should provide support out of considerations of prudence, but that, at least at the present time, society *owes* support to engineers. This conclusion follows from the fact that society is not neutral in the choice to become an engineer. Young people are encouraged, formally and informally, to pursue professional careers such as engineering. Having entered a university engineering program—usually without the knowledge or ability for much mature reflection—students are channeled into a rigorous engineering curriculum that usually offers them little idea of what to expect on the job while extolling the virtues of the profession. Society takes advantage of immature decisions, fosters only limited development of reflective abilities, and provides only selective knowledge of what work will be like. In such circumstances it is difficult for aspiring engineers to gain an adequate appreciation of the moral pressures that they will encounter, and so they do not have a proper opportunity to judge whether they are willing to take on the responsibility that goes with the job. Understanding one's moral predicament comes, if ever, further down the line when more substantial career and life commitments have been made and there is much more to lose in changing employers or careers. This is not a fair position for a person to be placed in and is, in fact, likely to be most trying for the morally conscientious engineer. In this situation, then, society not only ought to provide support out of considerations of prudence, but also is morally bound to provide at least some degree of assistance.

I do not, however, want to overestimate the degree to which engineering students are misled about the moral demands of the jobs they are training for. They have some responsibility to ascertain the facts for themselves. Enough stories of spectacular corruption and the tragedies of unforeseen consequences have been publicized to provide a general awareness of the potential for moral problems. However, I still find startling the credulity of many of my engineering students and attribute it, to a significant extent, to the limited and exploitative education they often receive.

There are a number of mechanisms of support for engineers which society may provide. These include effective government regulation, legal remedies, and the activities of professional organizations will to take strong measures, such as censure, boycott, and strike. It is not my purpose, however, to discuss these mechanisms at this time.[10]

I have claimed that engineers should be held to higher standards of care than others and argued that they should receive some measure of support from society in trying to meet those standards. To this I add the observation that not much effective support is presently available and there is not soon likely to be much more. This justifies sympathy for engineers in their predicament. It may even on occasion provide an excuse for certain sorts of complicity in wrongdoing. But it does not license total acquiescence to immorality. Nor do these allowances nullify the general requirement of a higher standard of care for engineers.

"If I don't do it, someone else will."

Next, a very brief comment on the defense, "If I don't do it, someone else will."[11] Obviously, the fact that someone else will perform an immoral action if one declines does not make the action right. The defense could be couched in other terms: "Why should *I* have to be the one to sacrifice when the bad consequences are inevitable anyhow?" One answer is, of course, that moral integrity requires it. Another reply is to challenge the inevitability of the worst consequences. Generally, one may comply grudgingly while doing the best one can to temper the bad consequences and to combat further activities of the same sort (rather than to promote immoral practices through cheerful complicity). When some improvement can be affected, grudging compliance may be the morally best path, but at other times disengagement may be the only morally acceptable course of action.

It is possible that one may be the victim of bad moral luck and may keep finding oneself in situations that demand sacrifice, but there are things that one can do to reduce the chances. Some of these will be mentioned in the course of my reply to the next defense.

"It's not my job."

This defense attempts to undercut the applicability of the principles I have stated. The claim here is that the engineer just does not have the power that I assert he has. According to this defense, the structure of corporations (as well as governments and other large heirarchical organizations) is such that people other than the practicing engineer make the critical decisions. Upper-level managers and executives are the ones who determine what projects shall be worked on, how work shall proceed, and what shall be done with the result. The engineer does not have and should not have discretionary power.[12] For example, engineers design an automobile within general guidelines that have been specified *for* them. The design they produce will have virtues and defects. They are responsible for pointing out the defects to the best of their abilities, but it is the managers and executives who decide whether or not to use the design and market the product. These individuals have control of technology, not the engineers.

The proper response to all this is to point out that discretionary power is not the only form of control. The harm that results from a dangerous product comes about not only through the decision to employ the design, but through the formulation and submission of the design in the first place. Harm could not come about if the engineers had refused to submit the design when they had good reason to believe that it was dangerous.

But now one might reply to this by acknowledging that the

engineer does have the *power* to forestall harm, but then claiming that it is not the engineer's *responsibility* to exercise this power. Indeed, it would be wrong for him to do so. A corporation is structured so that people perform various limited and specialized functions, thus taking advantage of special talents and defining (supposedly) clear lines of authority and responsibility. Engineers are trained, hired, and paid to do engineering work. Their responsibilities and prerogatives extend no further.

In response, I will allow that corporations can (morally) be structured in this way. But allowing this still does not relieve engineers either from moral responsibility for harm that may result from their work or from the moral necessity of their taking appropriate action on the basis of that responsibility. When one gets into a position in which one abdicates or delegates control over choice of work, how that work shall be done, what shall be done with it, and so on, it is morally incumbent on one to have good reason to believe that control will be exercised morally. When there is reason to believe that those who exercise the powers one has relinquished are not to be trusted to act morally, then one has a responsibility to press for recognition of moral values, to withhold one's contribution, or to sever one's relationship altogether, depending on the gravity of the case. Indeed, even where one's work does not directly contribute to harmful results, one may be subject to moral criticism for indirectly supporting an immoral organization.

There is an important implication of this argument for the criteria by which one should evaluate prospective employment. In considering employment one must weigh not only economic and career opportunities, but also the morality of the organization and the prospects for moral action within it. One must consider the morality of the structure and goals of the organization and of the particular individuals involved—colleagues, supervisors, managers, and executives. One must also reckon on changes that are likely to occur. Selecting a job merely on the basis of personal preferences for salary, location, potential for advancement, or challenge and interest of the work leaves one completely open to moral criticism. And this process of moral evaluation does not end at the point of employment. Re-evaluation of the potential for moral action in one's work is a continuing responsibility.

"There's no alternative."

Pointing out the responsibility of engineers to judge the morality of the organizations they enter leads to the final attempt I will consider to defend an engineer's morally questionable activities. The defense is this. What if there is reason to believe that virtually any employment will place one in a morally compromising position? Such a circumstance does not necessarily mean that all potential employers are evil. The situation may rather be the result of structural deficiencies in the corporate form of organization or of the pressures of competition on vulnerable people who have insufficient support. In such a situation, what is even the most moral and well-intentioned engineer to do? Having prepared in good faith for a career in engineering, is he now required by morality to abandon that career?

The answer, I think, is yes. Of course, it is a matter of degree. But in the extreme case the conclusion is clear: if one's only alternative for an engineering career is to design fire bombs for the destruction of London, then it's time to try to emigrate or to change career. The point is that the degree of one's commitment and the good faith with which one has dedicated one's talents do not change the morality of the case. One may, perhaps, blame others for misleading one about the moral situation, but this does not alter what is the moral course of action.

There is one other option. This is to accept employment, but to maintain a crusading attitude. This alternative can be morally acceptable if one can hope to have good effects while not contributing to greater evils.

The general conclusion I hope to have established, however, is that moral considerations must always be borne in mind. Engineers, managers, and the others within the corporate structure who develop and manage technology are never mere "animate machines" in the service of corporate ends.[13] Moral responsibility cannot be abdicated; corporate structure cannot shield one from responsibility as an autonomous moral agent.

NOTES

I would like to thank Mary G. Richardson and the editors of *Business and Professional Ethics Journal* for helpful comments.

This is a revised version of a paper presented at the Second National Conference on Ethics in Engineering in Chicago, in March, 1982. The conference was made possible by an EVIST grant from the National Science Foundation and was directed by Vivian Weil, Center for the Study of Ethics in the Professions, Illinois Institute of Technology.

1. For simplicity, I will usually refer hereafter only to (practicing) engineers and will consider specific objections with them primarily in mind.

2. A sample of codes of ethics in engineering can be found in Albert Flores, ed., *Ethical Problems in Engineering*, vol. 1, 2nd edition (Troy, NY: Rensselaer Polytechnic Institute, Center for the Study of the Human Dimensions of Science and Technology, 1980), pp. 63-75.

3. Most engineers practicing in the United States work for private corporations. Similar arguments to the ones I give can be applied to the public sector. For a treatment of moral responsibility in government, see Dennis Thompson, "Moral Responsibility of Public Officials: The Problem of Many Hands," *American Political Science Review* 74 (1980): 905-916.

4. In a sense that would constitute "strong role differentiation" as defined by Alan H. Goldman in *The Moral Foundations of Professional Ethics* (Totowa, NJ: Rowman and Littlefield, 1980), pp. 2-3.

5. Still further afield would be application of the principles to professionals (e.g., physicians, journalists) whose work is not in the development of technology.

6. Some of the main problems are raised crisply in Samuel Florman's *Existential Pleasure of Engineering* (New York: St. Martin's Press, 1976), chap. 3, especially pp. 21-22. I would not, however, endorse the solutions proposed there.

7. Readers unfamiliar with the philosophical literature may usefully consult *The Encyclopedia of Philosophy* (New York: Macmillan, 1967). See articles and bibliographies under such headings as "Golden Rule," "Problems of Ethics," "History of Ethics."

8. That other people's negligence or misconduct may be the critical factor through which harm results enormously complicates matters. I can offer only a brief comment. Take a case such as the illegal sale of hospital drugs by hospital personnel. Does this affect what a chemical engineer should do? To a certain extent the answer depends on the balance of good over evil that results from the engineer's work. True, other people are responsible for controlling illegal drug traffic, but, as I discuss below, morality requires that one consider and modify one's actions in light of the (fallible) workings of the *whole* system of which one is a part. In extreme cases, when great harm is likely to result, even though it be through other people's ignorance, negligence, or evil, one may have a moral responsibility not to continue. For example, I can see little moral justification for having *any* role in the rocketry work in Libya carried out by the German firm Otrag.

9. On culpability, see Kurt Baier, "Responsibility and Action," in *The Nature of Human Action*, ed. Myles Brand (Glenview, IL: Scott, Foresman, and Company, 1970), pp. 121-123.

10. A number of these mechanisms are discussed in Stephen H. Unger, *Controlling Technology: Ethics and the Responsible Engineer* (New York: Holt, Rinehart, and Winston, 1982), chap. 4, 5, and 6.

11. A number of perspectives on this defense in general are analyzed in the contributions of Jonathan Glover and M. Scott-Taggart to the symposium "It Makes No Difference Whether Or Not I Do It," *Proceedings of the Aristotelian Society*, suppl. vol. 49 (1975), pp. 171-209. See also section I of Michael Scriven, "Business Responsibilities in Product Design and Manufacture," in *Responsibilities in Product Design and Manufacture: Proceedings of the Second Panel Discussion of the Council of Better Business Bureaus* (Washington, DC: Council for Better Business Bureaus, 1978), pp. 92-103.

12. For a view of this sort, see Florman, especially chap. 3.

13. The "mechanical view" of corporate rationality is set out clearly in John Ladd, "Morality and the Ideal of Rationality in Formal Organizations," *Monist*, vol. 54 (1970), pp. 488-516. Ladd's article contains several useful references to the management literature on this topic.

ANDREW OLDENQUIST

Commentary*

This past summer, as I was boarding my DC-10 at Tokyo-Narita airport for the long flight to Kuala Lumpur, the smiling Japanese flight attendant said, "Have a nice fright." When I think of how many people, agencies, organizations, and companies are responsible for such frights (slight in my case, I admit), I cannot help but think that it is time the moral heat was taken off the engineers a little and distributed, without high indignation, among the many other places it also belongs.

Kenneth Alpern's thesis, that engineers are responsible for the foreseeable harm that may come from what they design, and to a degree proportionate to the gravity of the harm, is one with which I agree. There is no alternative to his conclusion that moral considerations must always be borne in mind, both in the appraisal of a design and before accepting employment. But I think he will make engineers resentful when he says they must be ready to "make greater personal sacrifices than can normally be demanded of other individuals," especially in the way he goes on about it later in his essay. It sounds, despite some disclaimers, as though engineers are singled out for potential martyrdom.

"I have claimed," he says, "that engineers should be held to higher standards of care than others and argued that they should receive some measure of support from society in trying to meet those standards." Higher than we require of others in their own work? Why? The answer is in his Corollary of Proportionate Care: "When one is in a position to contribute to greater harm or when one is in a position to play a more critical part in producing harm . . . , one must exercise greater care to avoid so doing."

Is a laborer less obliged to erect scaffolding safely and cover

* Reprinted with editorial changes by permission of the author. Copyright ©
1983 by Andrew Oldenquist.

dangerous holes? A bus driver may find the schedule dangerously fast; a pilot may notice countless dangers his airline condones; mechanics may be asked to perform maintenance below minimum safety standards, workmen told to make concrete with what they know is too much sand, corporate financial officers asked to go along with practices likely to ruin the company and its stockholders, secretaries expected to be blind to their swindling or embezzling bosses, and so on. And on the other side, engineers do lots of work in their daily routines with about the same potential for Great Harm as the work of the average receptionist.

"These principles do not single out engineers. The same principles and conclusion hold for anyone in a position of power—power to harm—from truck driver to president." But the tone of the essay makes engineers *sound* singled out, because the author appears to underappreciate the moral responsibility as well as the power of "nonprofessionals," such as those listed above, to do harm.

I think engineers, physicians, and other professionals are somewhat more strongly obliged than ordinary workers to exhibit integrity, trust, and self-policing, for the reason that society has given those with the credentials exclusive and privileged access to a type of work, and hence stewardship over it. Laborers aren't in that situation. But surely the laborer or mechanic *is* as obliged as the engineer to make things that won't collapse or break down; there is, for example, no cogent argument I know that lower socioeconomic status justifies diminished responsibility. In any case Alpern doesn't make the distinction between these two kinds of obligations.

Alpern says job hunters should appraise the morality of prospective employers. I agree, but without enthusiasm except in extreme cases. One *can* be a moral snob about these things. You should not become the sorcerer's apprentice or work for the Mafia. But should you refuse an offer from McDonnell Douglas because they have a bad record with one of their airplanes or from Dow chemical Co. because they made napalm? Possibly Alpern does not have that in mind but means you should refuse employment where the job you would personally do is morally suspect. The latter would occur very rarely. I suspect most refusals to solicit or accept a job on "moral grounds" are motivated by politics or political ideology.

We usually respect people's "decisions of conscience" not to design whaling equipment or a highway through an Indian reservation, but we should not confuse these sentiments with plain obligations, devoid of ideals and ideology, not to build scaffolding that is likely to collapse or market a drug likely to cause birth defects. It is not that Alpern says the wrong thing about this. Rather, he doesn't say anything about it, and I would like to know a little more about my obligation to continuously monitor for moral defects the structure, colleagues, bosses, goals, and possible future changes in my company, beyond being told that it is a matter of degree but that I shouldn't design fire bombs destined for London. Engineers are told not to do or even contribute to harm, with no specifics; does this mean engineers are to avoid evil, in whatever sense each individual puts on it? And, since there is some evil in every large enterprise, decide for oneself if the degree of it warrants opting out? That would be a relief; but engineers will worry that if they admit the principle of nonparticipation in what is "evil enough," someone will emerge with a list, a catechism.

Morality ought to require small to moderate concessions of self-interest, shared nearly equally by all, and should be clearly seen to make each of us better off, in general, than we would be without the system. If morality is made too expensive, or if it is given an uncompromising, terrible purity—a "though the heavens may fall" autonomy, a set of duties uninterested in human feedback—all but saints will reject it, and the rest of us will not call *that* morality. It is this tone of lofty, unyielding requirement that causes my only serious misgiving with the essay. And I am reasonably certain that if Alpern sat down and talked about cases and motives and human psychology, he would seem less like a marble statue.

But he says, "We have a basic obligation to avoid playing *any part* in the production of harm" (his emphasis). This may require total withdrawal from organized society: no payment of taxes, for they buy instruments of death; no work for any corporation or government, for they all do some harm. Or may we sum the harm and good a government or corporation does, and collaborate if the result is positive (as he seems to suggest in a footnote)? We are not explicitly told. Finally, we are offered the hypothetical case in which "virtually any employment will place one in a morally compromising situation." Ought I then abandon my engineering career? Yes, he says. But the terrible abstractness of this puts it in the same pigeonhole with "do good and avoid evil." The interpretations range from a priggish "You will be cooperating with misleading advertisers and with a subcontractor who intentionally underbid" to "The whole economy is making gas for extermination camps and weapons for world conquest." The principle itself warrants scarcely a thought until we begin the discussion of interpretations.

SAMUEL C. FLORMAN

Commentary[*]

When speculating about engineering ethics, professional philosophers have a way of conjuring up situations that illustrate a point but are, for the most part, the stuff of fiction. Alpern speaks of an engineer falsifying a report on safety tests. Engineers do not falsify reports, nor are they asked by their employers to falsify reports. It is pure Hollywood to envision young, idealistic engineers wondering whether or not to blow the whistle on corrupt corporate executives who would compromise public safety by producing a dangerous car/plane/dam/bridge/chemical—pick your own scenario. Or, to be more precise, in the extremely rare instances when an engineer encounters bad people engaged in deceptive practices—and it will not happen once in a thousand careers—he knows that he must not participate. It is unprofessional, immoral, and probably illegal. What could be more obvious? And, once having been said, what could be more trite?

Engineers occasionally do things that are injurious to the public. This is not because engineers are immoral, but because they are sometimes inaccurate, careless, or inattentive. I will not quarrel with Alpern's Principle of Care or even his Corollary of Proportionate Care. But from these I would conclude that engineers, in their work, must be accurate, careful, and exceptionally attentive. The essence of engineering ethics, I believe, is reliability. Since people are counting on them, engineers should do their darnedest not to make mistakes.

But even if engineers are honest and sincere (which they almost invariably are) and careful (which they usually are) this does not help us determine which products, structures, and systems best serve the interests of society. Alpern is concerned with the question of how we use technology and protect ourselves from the potential ill

---

[*] Reprinted with editorial changes by permission of the author. Copyright © 1983 by Samuel Florman.

effects of technology, and here in concentrating on engineering ethics he is very wide of the mark. We do not leave it to our soldiers to determine when we should have war or peace. Nor do we leave it to our judges to write our laws. Why, then, should we want our engineers to decide the uses to which we put our technology? Clearly we should not.

I was interested to see that in note 8 Alpern refers to rocketry work in Libya. It so happens that not long ago my company was invited to submit a bid on a building for the Libyan government in New York. As individuals—only incidentally as engineers—we did not care to work for a government that was said, among other things, to be assassinating disenchanted emigres in various parts of the world. This gave us a moment of righteous self-satisfaction. But would society's best interests be served if all engineers refused to work for Libya, even if the United States government approved of the building in question? I respect engineers who do not wish to work on weaponry—I personally prefer not to—but do we want engineers, as a group, to decide whether or not the nation should be armed? Are we pleased when the longshoremen's union refuses to load the ships of some nation whose politics they dislike? What if Alpern succeeded in politicizing engineers only to see them support causes alien to his philosophy—like the engineers in Ayn Rand's *Atlas Shrugged* who withdraw from society, in effect going on strike, until the government agrees to abandon "liberal" programs? Certainly we do not want engineers—even morally superior engineers—to make decisions that we have not designated them to make.

Nor should we expect engineers to define the risks that society takes in connection with technology. We live in a democracy, not a technocratic oligarchy. We the people decide what sort of risks we are willing to take. We do it imperfectly and with a great deal of confusion and debate, but we do it nevertheless. We do it through laws and regulations. And where these do not pertain, we do it through standards of accountability that are established by juries comprised of ordinary citizens. Alpern says, "Harm could not come about if . . . engineers . . . refused to submit the design when they had good reason to believe that it was dangerous." This ignores the fact that all designs contain some element of danger. In seeking to minimize danger, one usually increases cost, and therein lies a dilemma. In our society, low cost of production is considered desirable because we believe that people of modest means should have access to as many material benefits as possible. In attempting to make a product absolutely foolproof—a ladder that won't tip, for example, or a no-scald shower valve—we add cost that the average citizen may not want to pay in order to compensate for the carelessness of his neighbor. There is no great trick to making an automobile that is as strong as a tank and safe to ride in. The challenge is to make an automobile that ordinary people can afford and is as safe as the community thinks it should be. There are also considerations of style, economy of use, and effect on the environment.

When engineers and managers get together to design a product, they have a host of criteria to consider. It is simplistic to depict this complex process as a fight between good and evil in which the bad guys try to maximize profit by compromising on safety. First of all, reductions in manufacturing cost do not serve to swell profits. Reductions in cost serve to reduce the selling price. Profit most often results when a product is considered by the public to be "good value," and debate will rage within corporate headquarters concerning how best to capture a market, that is, how best to attract and satisfy the public.

Where risk to the public is entailed, people of good will can differ widely about standards that should be applied. In my daily work I am involved in construction of buildings, and in my office you will find a dozen engineers—all good people—with different ideas about what constitutes a "safe" building, about materials, exits, alarms, stairs, sprinklers, wind resistance, earthquake resistance, and on and on. Those who favor buildings that are less "safe" and more "economical" are not thinking of profit; they are thinking of conserved resources and affordable housing. I have heard hospital administrators decry safety requirements that they consider unnecessary—for example, a stair to be rebuilt six inches wider—arguing that costs added to health care put the entire community at higher risk.

Of course, all these individual opinions are beside the point because there are government standards and codes that determine what a safe building actually is. This is true, to the extent feasible, in all industries. It is the only process that makes sense.

There has been a steady growth of nationally accepted industrial standards—at present more than twenty thousand—and a proliferation of regulatory laws, agencies, and directives. This brings about complaints from people who wish that life was less complicated, but surveys show that this is the way that most citizens want to pursue. Regulations, codes, laws—made as sensible and streamlined as possible, but not at the cost of effectiveness—this is the best way to translate into action communal sentiments about technology.

Reliance upon the opinions—the bias!—of individual engineers in industry puts control of our destiny in the wrong hands. It is also a recipe for chaos. If in a corporate setting each of several hundred engineers was to come to work each morning prepared to argue for his or her vision of an appropriately designed product, no constructive work could be accomplished.

Increasingly in large corporations, product safety has become the province of designated specialists. These people rely upon scientific knowledge and are guided by codes and regulations that have been established by other designated specialists—usually in the public sector. Where codes and regulations do not apply, then they must consider standards of liability as defined by law and the courts. These standards, of course, are constantly changing, reflecting, or let us say approximating, the public will. Insurance companies and their specialists also are part of the equation, sensitive to the mood and expectations of the community.

It is not a confession of moral weakness to stress a reliance upon law. Law is a communal expression of moral choice. In considering how best to deal with technological risk, it is not appropriate to stress the personal preferences of the individual engineer.

I agree with Alpern that "moral considerations must also be borne in mind" and that "moral responsibility cannot be abdicated." But the moral responsibility of the engineer, I maintain, is to do good, careful work within the parameters established by the community, not to be guided by personal whim.

By all means, let engineers not do work that their conscience forbids them to do. And of course, like all other law-abiding citizens, let them shun evil and report misdeeds to the proper authorities. Within limits of conscience, however, each citizen may be called upon to participate in activities that he would rather see handled differently. We call this democracy.

By depicting technological design in terms of morality, Alpern does a disservice to his young students. They will find out soon enough, however, that engineering problems will not yield to good intentions. They will also find that many of their fellow citizens will blame technology for all sorts of ills while refusing to support the legislation, fund the agencies, or pay for the projects that would help cure these ills.

Ethical engineers, in addition to being careful, honest, and diligent in their work, will want to contribute time to the *pro bono* committees that set standards, investigate problems, and educate the public in technical matters. As citizens, they should also voice opinions and support causes in the political arena. These are better ways to do good, I believe, than undertaking the morally ambiguous role of vigilante in the workplace.

Select Bibliography

BOOKS

Baumrin, Bernard and Benjamin Freedman, eds., *Moral Responsibility and the Professions* (New York: Haven Publications, 1983).

Bayles, Michael D., *Professional Ethics* (Belmont, California: Wadsworth, 1981).

Burnham, Frank, *Cleared to Land—The FAA Story* (Fallbrook, California: Aero Publishers, 1977).

Eddy, Paul, Elaine Potter, and Bruce Page, *Destination Disaster* (New York: New York Times Book Co., 1976).

Ewing, David W., *Freedom Inside the Organization: Bringing Civil Liberties to the Workplace* (New York: Dutton, 1977).

Flores, Albert, ed., *Ethics and Risk Management in Engineering* (New York: University Press of America, 1989).

Flores, Albert, *Professional Ideals* (Belmont, California: Wadsworth, 1988).

Frank, Nancy and Michael Lombness, *Controlling Corporate Illegality* (Cincinnati, Ohio: Anderson Publishing Co., 1988).

Godson, John, *The Rise and Fall of the DC-10* (New York: David McKay Company, 1975).

Iannone, Pablo, ed., *Contemporary Moral Controversies in Technology* (Oxford: Oxford University Press, 1987).

Johnson, Deborah, *Ethical Issues in Engineering* (Englewood Cliffs, New Jersey: Prentice Hall, 1991).

Johnson, Moira, *The Last Nine Minutes* (New York: William Morrow, 1976).

Kemper, John D., *Engineers and Their Profession*, 4th ed. (Philadelphia: Saunders College Publishing, 1990).

Lombardi, Louis, *Moral Analysis* (Albany, New York: SUNY Press, 1988).

McClement, Fred, *Jet Roulette: Flying Is a Game of Chance* (New York: Doubleday, 1978).

Martin, Mike W. and Roland Schinzinger, *Ethics in Engineering* (New York: McGraw-Hill, 1983).

Nader, Ralph, Peter Petkas, and Kate Blackwell, eds., *Whistleblowing: The Report of the Conference on Professional Responsibility* (New York: Grossman Publishers, 1972).

Nance, John J., *Blind Trust* (New York: Quill/William Morrow, 1986).

Newhouse, John, *The Sporty Game* (New York: Alfred A. Knopf, 1982).

Noble, David, *America By Design: Science, Technology and the Rise of Corporate Capitalism* (New York: Alfred A. Knopf, 1977).

Norris, William, *The Unsafe Sky: The Unvarnished Truth about Air Safety* (New York: W. W. Norton, 1981).

Perrow, Charles, *Normal Accidents: Living with High-Risk Technologies* (New York: Basic Books, 1984).

Perrow, Charles, *Complex Organizations*, 2nd ed. (Glenview, Illinois: Scott, Foresman and Company, 1979).

Petroski, Henry, *To Engineer Is Human* (New York: St. Martin's Press, 1982).

Ramsden, J. M., *The Safe Airline* (London: Macdonald and Jane's, 1976).

Schaub, James, *Engineering Professionalism and Ethics* (New York: John Wiley and Sons, 1983).

Stewart, Stanley, *Air Disasters* (London: Arrow Books Ltd., 1986).

Unger, Stephen, *Controlling Technology* (New York: Holt, Rinehart and Winston, 1982).

Westin, Alan, ed., *Whistle-blowing!* (New York: McGraw-Hill, 1981).

Winner, Langdon, *Autonomous Technology* (Cambridge, Massachusetts: MIT Press, 1977).

ARTICLES

Blumberg, Phillip, "Corporate Responsibility and the Employee's Duty of Loyalty and Obedience: A Preliminary Inquiry," *Oklahoma Law Review* 24 (1971): 279-318.

Chalk, Andrew, "Market Forces and Aircraft Safety: The Case of the DC-10," *Economic Inquiry* 24, no. 1 (1986): 43-60.

Conway, John H., "Protecting the Private Sector At Will, Employee Who 'Blows the Whistle': A Cause of Action Based on Determinants of Public Policy," *Wisconsin Law Review*, no. 3 (1977): pp. 777-812.

Dickie, Robert B. and Robert W. Orlando, "Federal Response to the American Airlines DC-10 Crash—A Study in the Regulatory Process," *Journal of Contemporary Business* 9 (1980): 59-67.

"McDonnell Douglas' Billion-Dollar Gamble," *Forbes*, August 1, 1969.

May, Larry, "Vicarious Agency and Corporate Responsibility," *Philosophical Studies* 43 (1983): 69-82.

Thompson, Paul B., "Risking or Being Willing: Hamlet and the DC-10," *Journal of Value Inquiry* 19 (1985): 301-310.

Wise, T. A., "How McDonnell Won Douglas," *Fortune*, March 1967.

GOVERNMENT DOCUMENTS

Air Safety: Selected Review of FAA Performance, Report by the Special Subcommittee on Investigations of the Committee on Interstate and Foreign Commerce, House of Representatives, Ninety-Third Congress, December 1974.

Airworthiness of the DC-10, Hearing before the Subcommittee on Aviation of the Committee on Commerce, One Hundred and First Congress, Senate, September 19, 1989.

Aviation Safety: Management Improvement Needed in FAA's Airworthiness Directive Program, General Accounting Office, February 1990.

DC-10 Certification and Inspection Process, Hearings before the Subcommittee on Aviation of the Committee on Commerce, Science, and Transportation, Senate, Ninety-Sixth Congress, July 11 and 12, 1979.

DC-10 Engine Failure/FAA R and D Needs, Hearing before the Subcommittee on Transportation, Aviation and Materials of the Committee on Interstate and Foreign Commerce, House of Representatives, One Hundred and First Congress, November 9, 1989.

Design Analysis of Wide-Body Aircraft, Hearings before the Subcommittee on Investigations and Oversight of the Committee on Science and Technology, House of Representatives, Ninety-Sixth Congress, July 17, 18, 21; August 6, 15; October 4, 1979.

FAA Certification Process, Hearings before the Subcommittee of the Committee on Government Operations, House of Representatives, Ninety-Sixth Congress, June 11, 18; October 9 and 10, 1979.

Final Report, Secretariat of State for Transport, Accident to Turkish Airlines DC-10 TC-JAV in the Ermonville Forest on 3 March 1974 (Paris).

National Transportation Safety Board Report, NTSB-AAR-73-2 (Windsor).

National Transportation Safety Board Report, NTSB-AAR-79-17 (Chicago).

National Transportation Safety Board Report, NTSB-AAR-90-06 (Sioux City).

Oversight Hearings on the DC-10 Aircraft, Hearings before the Subcommittee on Aviation of the Committee on Commerce, Senate, Ninety-Third Congress, March 26 and 27, 1974.

Report on the Oversight Hearings and Investigation of the DC-10 Aircraft, Committee on Commerce, Senate, Ninety-third Congress, June 1974.

Review of Procedures and Policies of the FAA and NTSB with Respect to the DC-10 Cargo Doors, Hearings before the Special Subcommittee on Investigations of the Committee on Interstate and Foreign Commerce, House of Representatives, Ninety-Third Congress, March 27 and April 9, 1974.

Safe Skies for Tomorrow: Aviation Safety in a Competitive Environment, U.S. Congress, Office of Technology Assessment (Washington, DC: U.S. Government Printing Office, July 1988).

To Enhance the Safety Mission of the FAA, Hearings before the Subcommittee on Aviation of the Committee on Public Works and Transportation, House of Representatives, Ninety-Sixth Congress, August 20 and 21, 1980.

Institute of Electrical and Electronics Engineers Code of Ethics*

We, the members of the IEE, in recognition of the importance of our technologies in affecting the quality of life throughout the world, and in accepting a personal obligation to our profession, its members and the communities we serve, do hereby commit ourselves to the highest ethical and professional conduct and agree:

1. to accept responsibility in making engineering decisions consistent with the safety, health and welfare of the public, and to disclose promptly factors that might endanger the public or the environment;
2. to avoid real or perceived conflicts of interest whenever possible, and to disclose them to affected parties when they do exist;
3. to be honest and realistic in stating claims or estimates based on available data;
4. to reject bribery in all its forms;
5. to improve the understanding of technology, its appropriate application, and potential consequences;
6. to maintain and improve our technical competence and to undertake technological tasks for others only if qualified by training or experience, or after full disclosure of pertinent limitations;
7. to seek, accept, and offer honest criticism of technical work, to acknowledge and correct errors, and to credit properly the contributions of others;
8. to treat fairly all persons regardless of such factors as race, religion, gender, disability, age, or national origin;

---

* Reprinted, with permission, from the Institute of Electrical and Electronics Engineers. Copyright © 1990 by IEEE.

9. to avoid injuring others, their property, reputation, or employment by false or malicious action;
10. to assist colleagues and co-workers in their professional development and to support them in following this code of ethics.

Index

Made in the USA
Lexington, KY
01 October 2014